Charles J. Byrne

The Far Side of the Moon
A Photographic Guide

 Includes CD-ROM

 Springer

Charles J. Byrne
Image Again
Middletown, NJ
USA

ISBN-13: 978-0-387-73205-3   e-ISBN-13: 978-0-387-73206-0

Library of Congress Control Number: 2007932624

Printed on acid-free paper.

9 8 7 6 5 4 3 2 1

springer.com

Preface

The nature of the far side of the Moon has always been a mystery. Throughout human history, the far side of the Moon has been so near, its presence so obvious, and yet so hidden. It was only in the decade of the 1960s that we have been able to observe it, as opposed to inferring its existence. I have been fortunate to live during the period when the far side became visible and has been extensively examined by means that were not even known 50 years ago. During this time, we have learned much but of course we have asked new questions about secrets still hidden on the Moon. Some of these secrets will be revealed in the course of future exploration and robotic spacecraft missions. This book is about what we now know of the far side of the Moon, about what the current questions are, and about some possible answers.

It is a companion book to *The Lunar Orbiter Photographic Atlas of the Near Side of the Moon*, my first book (Byrne, 2005), also published by Springer in 2005. Together, these books cover the entire Moon in photographs taken by the robotic spacecraft that have done much of their work behind the Moon, out of communication with Earth for half of their orbital periods.

The photographs of the near side book were taken by a single mission of the Lunar Orbiter project, Lunar Orbiter 4 that was flown in 1967. This book is primarily composed of photographs from each of the Lunar Orbiter missions of the National Aeronautics and Space Administration (NASA) and from the Clementine mission of the Naval Research Laboratory (NRL). In addition, it contains images from the Russian Luna 3 and Zond 3, NASA's Apollo 16, and Japan's Nozomi.

The Clementine mission provided a comprehensive survey of altitude, albedo (intrinsic brightness), and multispectral data in 1994. Lunar Prospector provided gamma ray spectroscopy in 1998. The data from these spacecraft have added insight into the mineral composition of the lunar surface (supported by ground truth from analysis of lunar rocks and soil returned by Apollo and Luna missions). Clementine imagery is used extensively in this book where it is superior in topographic resolution or uniformity of coverage.

The reason for using photos from so many missions is that there has not been a single mission to systematically take pictures to show the topography of the far side of the Moon. Yet it has been possible to provide complete coverage of the far side by assembling photos from many sources.

Many years ago, I had the pleasure of working on the Lunar Orbiter project, representing the Apollo program in its search for safe and interesting landing sites. I worked for Bellcomm, a contractor to NASA headquarters. The Lunar Orbiter project was amazingly successful for the time, providing extensive new images in five out of five missions. This performance is a credit to Boeing, the prime contractor and manufacturer of the spacecraft, to Eastman Kodak, who made the camera system, and to the project management team at NASA's Langley Research Center, directed by Lee Scherer of NASA headquarters. The Langley team, who included Cliff Nelson, Israel Taback, and Norm Craybill, later managed the Viking project, which sent the first lander to Mars.

The limited scanning technology of the time resulted in artifacts in the images that distract a viewer. There are bright lines running across the mosaics between framelets and brightness variations from the spacecraft's scanner that appear as streaks within the framelets. Lunar scientists have become used to these artifacts, but they detract from the value to students and casual observers.

Since the first photos were received, I have wanted to clean up the scanning artifacts. Advances in the art of computation and the capacity of modern computers has enabled processing of the photos to remove nearly all of the scanning artifacts, resulting in clear images that are much easier to view.

Drawing on an understanding of the nature of the artifacts, I have written a software program that measures and compensates for the systematic artifacts and designed a filter that further reduces them, with minimal impact on the images of the Moon. This process is described in Appendix A.

It was with great satisfaction that, with the publication of this book, I have completed the publication of a comprehensive set of cleaned photographs from the Lunar Orbiter project. The photos are those that were selected by D. E. Bowker and J. K. Hughes for their book *Lunar Orbiter Photographic Atlas of the Moon* (Bowker, 1971). All of the far side photographs of Lunar Orbiter and the other photographs in the book are on the enclosed Compact Disc. A CD with the near side photographs was provided with my near side book.

While preparing these two books, I attended a series of yearly meetings of the Lunar and Planetary Science Conference (LPSC), sponsored by the Lunar and Planetary Institute, a NASA contractor. In the course of these meetings, where lunar and planetary scientists report on their current work, I became interested in the differences between the near and far sides of the Moon, and the attempts to explain that difference. Consequently, while preparing for this book on the far side, I took some time off to see if I could contribute to this question. It turned out that my background in communication research, specifically the art of finding signals in noise, led to a new hypothesis, with supporting quantitative evidence. I took the approach that a giant basin on the near side would be a signal, a pattern of elevation data, and that the subsequent history of bombardment of the Moon would be noise, random perturbations of the signal. Just as noise adds to a signal in communications, the principle of superposition establishes that the characteristic form of an impact adds to the form of an earlier basin, especially one of a much larger scale.

Accordingly, I modeled the elevation signal that would be produced by a giant basin, and varied the characteristics of such a basin, much like tuning a radio, until the maximum signal-to-noise ratio was found. Then the Near Side Megabasin revealed itself. This resulted in the parameters of the Near Side Megabasin that I reported to LPSC 2006 (Byrne, 2006). The method of this analysis is described in Chap. 13.

It remained to explain the crustal thickness data, which was qualitatively in agreement with the Near Side Megabasin, but in quantitative disagreement with the parameters I had first estimated (Byrne, 2006) by nearly an order of magnitude. Fortunately, I was able to contact H. Hikida and M. A. Wieczorek, who are actively studying the crustal thickness problem (Hikida, 2007), and they generously shared their new data. This reinforced the qualitative agreement with the ejecta that would be generated by the Near Side Megabasin, but also reiterated the quantitative disagreement.

The point that I had missed in 2006 was that the Moon was soft and easily deformed in the early times when the megabasins were formed. This has long been known (Lemoine, 1997) but not by me. The great weight of the ejecta on the far side would have caused it to sink, retaining its shape, but not its scale. This phenomenon is known as isostatic compensation. Given the accepted densities of the mantle and crust, isostatic compensation would cause the ejecta of the Near Side Megabasin to sink into the mantle, leaving only one-sixth of the initial depth of the deposit, accounting for the quantitative discrepancy.

The implication of the crustal thickness data is that both the cavity of the Near Side Megabasin and its ejecta would have been six times deeper at the time of its formation, before isostatic compensation, than the current topography would require.

This new modification to the Near Side Megabasin hypothesis is in quantitative agreement with all the relevant data that I have been able to review, and it answers many questions about the topography and crustal thickness of the Moon. I hope that it will be a lasting contribution to the understanding of the Moon, and that it will stimulate further questions and topics of research.

Several sources contributed to the digital imagery in this book. Lunar Orbiter photography has been archived as hard copy photographs, each about 60 cm (about 2 ft) wide, at each NASA Regional Planetary Image Facility, including one at LPI in Houston, Texas. LPI has digitized this important archival source and published the images in the Digital Lunar Orbiter Photographic Atlas of the Moon on the LPI Web site (http://www.lpi.usra.edu). Further, the LPI staff has added annotations to the photos, outlining nearly all of the named features.

A team led by Jeff Gillis carried out this important work (Gillis, 2002); Jeff was supported by Washington University at St. Louis and is currently with the University of Hawaii. LPI technical and administrative support was provided by Michael S. O'Dell, Debra Rueb, Mary Ann Hager, and James A. Cowan with assistance from Sandra Cherry, Mary Cloud, Renee Dotson, Kin Leung, Jackie Lyon, Mary Noel, Barbara Parnell, and Heather Scott. The selection of photos that were made available on the LPI Digital Archive is that selected for *Lunar Orbiter Photographic Atlas of the Moon* by Bowker and Hughes (Bowker, 1971).

NASA, NRL, the Russian space agency, and the Japan Aerospace Exploration Agency (JAXA) have made public the base data for the photos in this book. Processed images were downloaded from web sites at NRL, the United States Geological Survey (USGS), and the Jet Propulsion Laboratory (JPL). Ricardo Nunes and Phil Stookes have contributed mosaics.

Photos with annotated overlays identify the major features within each of the photos. These overlays were extracted digitally from those published by LPI, but I made additions, especially for the Clementine images, in the style established by LPI.

The names of features, both by LPI and myself, are aligned with the list approved by the International Astronomical Union (IAU), as maintained by Jennifer Blue of the USGS Astrogeology Branch. Notes with each photo point out salient aspects of the features. These notes, as well as those for my near side book, have been reviewed by Don Wilhelms, USGS retired, the author of my primary reference, who has taught me a great deal about lunar geology. The assignments of geological age to many far side features are derived from his book, which has contributions by John McClauley and Newell Trask (Wilhelms 1987). A reference that I used heavily in the preparation of this book is *The Clementine Atlas of the Moon* by Ben Bussey and Paul Spudis (Bussey, 2004). The dimensions of features (except for basins) are taken from this book, which in turn is based on the IAU catalog at USGS. Of course, I am personally responsible for any errors in this final book.

The combination of cleaned photos, labeled features, and notes are intended to serve as powerful aids to learning the geography of the Moon as well as valuable reference material.

My love and gratitude go to my wife Mary, who read and re-read the early drafts of this book and made many helpful comments.

Throughout the project of cleaning the photos and writing this book helpful suggestions and comments were made by Tammy Becker, Jeff Gillis, Mary Ann Hager, Ray Hawke, Paul Spudis, Ian Garrick-Bethell, Mark Robinson, Peter Schultz, Ewen Whitaker, and Don Wilhelms.

Charles J. Byrne

Table of Contents

Chapter 1

The Far Side of the Moon

1.1. The Far Side Unknown: A Mystery

The far side of the Moon is the side that we cannot see from Earth. Because the rotation of the Moon is locked in synchronism as it revolves around Earth, its near side always faces Earth. If people lived on the Moon, those who lived on the near side would always see Earth and those who lived on the far side would never see it.

This situation (which is common to the large moons of all planets) has led to an aura of mystery about the far side of the Moon ever since humans thought about the night's illumination. When humanity thought it was at the center of the universe, this condition seemed natural enough, but when the Copernican view of the solar system prevailed, there was a period when people wondered what the far side might be like.

Was there a pattern there like the one on the near side? Was it completely different? Was it colored, like Mars, or nearly colorless like the near side? Were there seas and oceans, as once were thought to be on the near side?

The near side of the Moon is a symbol of romance. The far side has often been called the dark side of the Moon, as if our inability to see it meant that it was never lit (it receives as much sunlight as the near side). As an example of our feelings, in the enormously popular album "The Dark Side of the Moon" by Pink Floyd, the far side of the Moon is a symbol for depression, for alienation.

When space probes Luna 3, Zond 3, and Lunar Orbiter 1 transmitted images back to Earth (Figure 1.1), it became apparent that the nature of the far side of the Moon was indeed very different from that of the near side. Many questions were answered, but many new ones were raised.

1.2. The Far Side Revealed: A New Mystery

Surprisingly, there were only a few patches of mare on the far side. There was no "Woman in the Moon" to travel with the "Man in the Moon." It was more like we were seeing the back of the head of the Man in the Moon. The far side of the Moon is remarkably uniform at first view, heavily cratered and with little variation in brightness (Figures 1.2 and 1.3).

Even before the Clementine mission, examination of the Lunar Orbiter photography identified a number of large basins on the far side including the giant South Pole-Aitken Basin. Many of these basins are outlined in Figure 1.4, and are shown in great detail in photographs and images of this book. Now we knew more about both similarities and differences between the near side and the far side. Mysteries were resolved, but the new mystery was how the striking differences in the two sides of the Moon came about.

The differences are so great that Robert Lepage used the contrast as a metaphor in his film "Far Side of the Moon" for extreme differences between personalities and to the competitive space race between the United States and the Soviet Union.

Since the polar orbit was oriented so that the sun was directly behind Clementine as it passed over the Moon's equator, the images show the inherent brightness (albedo) of the Moon. Shadows appear only near the poles. Figures 1.2 and 1.3 are lambert equal-area projections.

1.3. The Current View of the Far Side

This book provides comprehensive coverage of the far side of the Moon, the only book that has been dedicated to the far side alone. It is a guide to the different regions of the far side and the interactions among the features there. Photographs from the five Lunar Orbiter missions provide most of the coverage in this book. The photographs presented here have been computer-processed to remove the artifacts introduced in scanning them in the spacecraft and reproducing them on Earth and are cleaner and easier to understand than the photos that have been used in research for the past 40 years. The cleaning process is described in Appendix A.

Although the primary purpose of the five Lunar Orbiter missions was to survey the near side in preparation for the Apollo landings, they covered all of the far side as well. Photographs from several other space missions are included, especially where they provide coverage that complements that of the Lunar Orbiter missions. In particular, photos from the Clementine missions cover the Polar Regions.

Figure 1.1. The western far side of the Moon looking back to Earth. This picture is a mosaic made from Lunar Orbiter photos LO1-116M and LO1-117M, taken in August, 1966. The photos have been cleaned of artifacts as described in Appendix A. NASA, LPI, photos cleaned and assembled by the author.

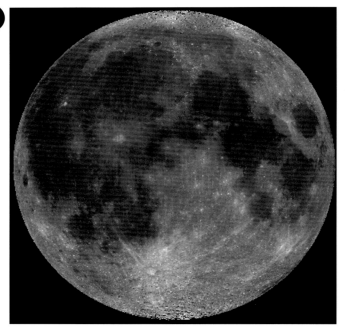

Figure 1.2. The Clementine spacecraft gave us full coverage of the Moon by looking vertically down from a polar orbit. This is the familiar near side, at a wavelength of 750 nm wavelength. NASA, LPI (Spudis).

Figure 1.3. The far side is relatively uniform.

The Clementine mission covered the entire Moon systematically, but its primary purpose was to provide multispectral imagery at minimal shadowing. Consequently, there is little information about the shape of the surface except near the poles, where the sun angle is more favorable to showing the topography and the images are superior to the Lunar Orbiter photos.

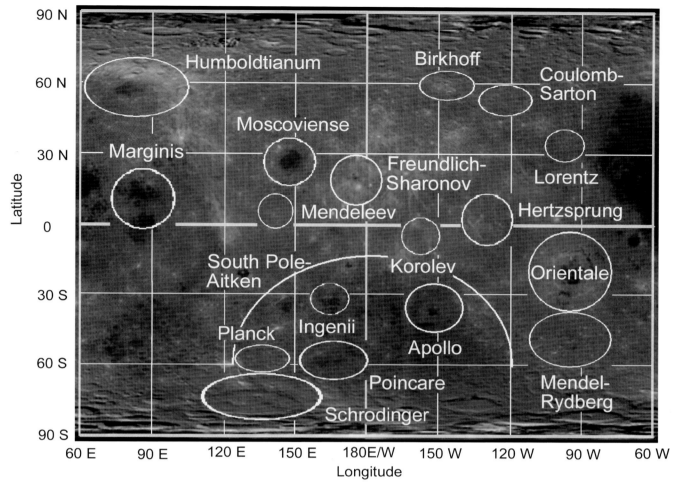

Figure 1.4. The Clementine view of the far side of the Moon in a geometric projection (a graph of latitude against longitude). Many of the basins are shown here to provide reference points. The basins near the poles appear elongated by the stretching of the projection. NRL base map.

The Apollo missions also had a subordinate mission to photograph the far side. Because of their equatorial orbits and lighting conditions, their excellent photographs are limited to the equatorial region from the central near side, around the eastern limb, to the central far side. Consequently, only a few of the Apollo photographs appear here.

The far side is divided into regions based on the features that have influenced them. The form of presentation changes in each chapter to accommodate the nature of the photography since it is derived from a number of different missions.

Notes describe interesting aspects of each image, explaining the geologic nature and relative ages of the features, with special attention to the nature of their interactions. The locations and sizes of features whose names are recognized by the International Astronomical Union are shown.

A compact disk (CD) is enclosed. All of the photographs in this book are included in this CD in JPEG format. They can be viewed on monitors, printed, or used for illustrations in talks and articles.

1.4. Mystery Resolved? The Near Side Megabasin

The main purpose of this book is to provide comprehensive images of the far side of the Moon produced by the Lunar Orbiter, Clementine, and Apollo missions. But in organizing the material, my interest turned to seeking an explanation for the dichotic nature of the Moon. Is there an explanation for the dramatic differences between the near and far sides?

Astrophysicists and astronomers have been puzzling over this question ever since the evidence arrived. An early hypothesis, first put forward by Wood in 1973 (Wood, 1973), is that a large impact on the near side has thrown its ejecta onto the far side. Since then, Cadogan (Cadogan, 1974) suggested the Gargantuan Basin, Whitaker (Whitaker, 1981) proposed the even larger Procellarum Basin, and Lucey suggested a basin centered in the southeast of the near side (Feldman, 2002). None of these hypotheses were accompanied by convincing topographic evidence, and in time, all proposals, while not completely refuted, where not confirmed either.

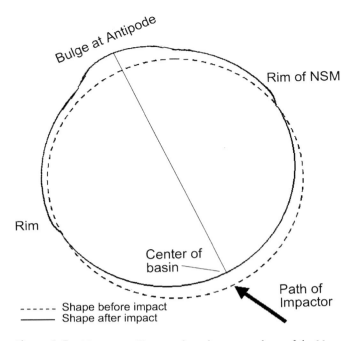

Bulge at Antipode

Rim of NSM

Rim

Center of basin

Path of Impactor

- - - - - Shape before impact
———— Shape after impact

Figure 1.5. This cartoon illustrates how the current shape of the Moon may have come about. It is suggested (Byrne, 2006, 2007) that a very large comet or asteroid (perhaps 600 km in diameter or more) struck the Moon at the point that is now at 8.5° north latitude and 22° east longitude. The impact excavated a cavity that covers more than half the Moon, nearly all of the near side except a small area near the southwest limb. The ejecta concentrated at the antipode of the point of impact, forming a large bulge. The elevations in this figure are exaggerated. See Chapter 13 for more detail on the Near Side Megabasin.

My personal search for a large near side basin found a site that had not previously been considered, and a size that far exceeded previous proposals. A model of this feature, scaled up from the shapes of lunar basins, fit the evidence of the Moon's unusual form, as measured by the laser-ranging LIDAR instrument carried by Clementine. This hypothetical impact, which I have informally called the Near Side Megabasin (Figure 1.5), provides a quantitatively supported explanation of the differences between the near and far sides of the Moon. The supporting evidence for the Near Side Megabasin is provided in Chapter 13, at the end of this book. Its center is to the northeast of the center of the Moon as we see it, at 8.5° north latitude and 22° east longitude. The interior of this basin covers more than half of the Moon; its rim is nearly all over on the far side. It has covered all of the far side with its ejecta, creating a mound that is currently about 5 km thick in the region opposite to the Near Side Megabasin's center (its antipode). In short, this feature apparently dominated the nature of both the entire far side and the entire near side and is a potential explanation for the major differences between the two.

Throughout this book, the major influence of the Near Side Megabasin is mentioned in the notes on the photographs and marked on some of the photos. If this feature is confirmed by further examination, this book will serve as an early documentation of it. Should it be refuted, then the mystery will remain but this book will still serve as a comprehensive photographic guide to all other features on the far side of the Moon.

Chapter 2
The Spacecraft Missions and Images

2.1. Spacecraft Programs

This chapter describes the programs that have contributed photography to this book on the far side of the Moon. It does not discuss the missions that have been limited to the near side, where much has been learned in detailed examination of the surface and sample return. The missions are described in chronological order.

Luna 3 and Zond 3

The former Soviet Union launched two flybys of the Moon: Luna 3 in October, 1959 and Zond 3 in July, 1965. Luna 3 covered the western part of the far side and Zond 3 covered the eastern part of the far side (beyond the western limb, as we see it from Earth). Between them, they photographed most of the far side.

Lunar Orbiter

The series of five NASA Lunar Orbiter missions, starting with Lunar Orbiter I in August, 1966, provided improved resolution and picture quality, covering nearly the entire Moon. The photos these spacecraft returned showed us the topography of the Moon and provided the primary observations for the comprehensive geologic analysis performed by the USGS Astrogeology Branch and others. This photographic record is still the starting point for research on the topographic features of the Moon. Lunar Orbiter I took the famous picture of the Earth, looking back from beyond the Moon; in a sense, this was the first self-portrait of our planet. Most of the photographs in this book are from the Lunar Orbiter missions. Each of the missions has provided some of the total coverage of the far side of the Moon.

Apollo

The Apollo program made many contributions to our understanding of the Moon, including the photographic coverage taken from lunar orbit by cameras in the Command Module. They provided the first photographic negatives returned to Earth, free of the artifacts introduced by transmission from the Moon. Because of Apollo mission's constraints, these photos were limited in their coverage to a band near the lunar equator. The far side photos were further limited by the lighting conditions to about 110° from the East Limb of the Moon. The first color pictures of the Earth from beyond the Moon were taken by Apollo 10, a manned mission into lunar orbit that was a rehearsal for the series of landing missions begun by Apollo 11 in July, 1969. In the later Apollo missions (15, 16, and 17) panoramic and mapping cameras joined the hand-held Hasselblad cameras to provide more systematic coverage, within the mission constraints. Apollo photographs in this book provide broad coverage of parts of the East Limb.

Clementine

In 1994, after more than 20 years of analyzing the results from the earlier mission, the Naval Research Laboratory launched the Clementine spacecraft. This spacecraft entered a near-polar orbit, and while the Moon rotated under its orbital plane, achieved uniform coverage of the entire Moon. This was the first mission to cover the Moon in a way that provided "the big picture." The images showed the Moon in a way that permitted direct comparison of the near and far sides; until then, equivalent views were manually constructed by airbrushed representations of rectified photographs. In addition to the visual imagery, Clementine carried multispectral sensors that provided data about mineral composition of the surface. A laser altimeter (LIDAR) carried by Clementine provided an accurate measurement of the elevations of the Moon's surface, at a resolution of about 0.25° (about 7.5 km). Clementine images are used in this book for the polar sections and broad coverage of the South Pole-Aitken Basin.

Lunar Prospector

In 1998, NASA launched the Lunar Prospector spacecraft, which introduced a novel instrument that measured the energy spectrum of solar neutrons that rebounded from the lunar surface. Comparison of the spectrum of the rebounding neutrons from the Moon with the spectrum of neutrons directly from the sun showed the distribution of energy lost from the collisions with the nuclei of the elements on the surface, and therefore their atomic weight. The resulting distribution of thorium was a clue in the development of the Near Side Megabasin proposal. Although results from Lunar Prospector are discussed, this book does not contain images that are derived from the Lunar Prospector instruments, because they are best presented in false color (Prettyman, 2007).

Nozomi

Nozomi was launched toward Mars on July 3, 1998, by the Japan Aerospace Exploration Agency (JAXA). On its way, looking back toward Earth, it took a picture of the far side of the Moon. Unfortunately, the maneuver to insert Nozomi into Mars orbit was unsuccessful, but the project achieved a unique view of the Moon.

2.2. The Images

It is not so easy to obtain imagery from space, mostly with highly automated spacecraft that encounter an environment that is very unlike that on Earth, but must return what they sense to earth for interpretation by scientists. Each program solves its problems in its own way, accomplishing the goals of its investigators using the technology available to its designers. Keep in mind that the robotic spacecraft collected all of the imagery of the far side of the Moon while out of communication with Earth. In the case of some of the missions, human operators had extensive control from orbit to orbit, but the spacecraft and their instruments were on their own when they were behind the Moon.

With the historic record available to us, we are free to integrate the products of these missions for our own goals, to make them available to more people.

With that background in mind, let us survey the imagery available to form a view of the far side of the Moon.

Luna 3 and Zond 3

The Luna 3 and Zond 3 spacecraft used vidicon cameras to take single frames of images, store them in memory during their flybys, and transmit them to Earth. Expecting a pattern like that of the front side of the Moon, and needing bright illumination, the mission planners scheduled their flybys when the sun would be high. Luna 3 continued on past the Moon and transmitted as much as it could of its memory at long range. Zond 3 returned toward (but not to) Earth and had better communications geometry.

Ricardo Nunes, an amateur astronomer in Portugal, created mosaics of individual images from these two spacecraft. (see Figures 2.1 and 2.2). Both mosaics have been produced by image processing techniques developed in our own times, long after the images were obtained and examined.

Lunar Orbiter

The Lunar Orbiter spacecraft provided two innovations for lunar imagery. First, they carried retro-rockets to inject them into orbits around the Moon, just as spacecraft orbited earth. This allowed for an extended mission, so that the lighting could be optimal for a large range of lunar longitudes. Second, they used photographic film to record the images. The negatives were developed on the spacecraft, using a Bimat process whereby the negative was placed in contact with a second emulsion carrying developing solution. The developed negatives were scanned by a combination of mechanical and electronic techniques, transmitted to Earth, and the image was reconstructed there. This permitted a combination of high resolution with broad coverage, far more than vidicon tubes could provide, and even more than today's electronic cameras that use CCD technology.

Each Lunar Orbiter exposure consisted of two frames taken by different lenses in two positions on the same strip of film. One image was called high-resolution (limited cov-

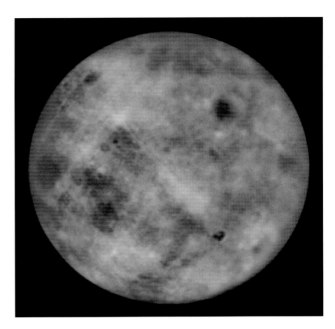

Figure 2.1. Mosaic of the western far side of the Moon and the East Limb, taken by Luna 3. The dark spot in the upper right has been named Mare Moscoviense. The small spot directly below it is mare in the floor of the crater Tsiolkovskiy. The dark areas to the left of Tsiokovskiy are Mare Marginis and Mare Smythii, both centered on the near side but straddling the limb. Image processing and mosaic of Figures 2.1 and 2.2 by Ricardo Nunes.

Figure 2.2. Mosaic of the eastern far side of the Moon and the West Limb, taken by Zond 3. This processed image is centered on the Orientale Basin, with Mare Orientale within it. The dark spot to the right of Mare Orientale is mare material in the floor of the Grimaldi Basin. The gray area in an arc along the right edge of the Moon is Oceanus Procellarum.

erage) and the other medium-resolution (wider coverage) (Figure 2.3). The relation between the two frames is shown in Figure 2.4.

Depending on the orientation of the spacecraft, the H1 frame can be either to the north or to the south, as shown.

Figure 2.3. Image of the Orientale Basin, a mosaic composed of two Lunar Orbiter medium-resolution photographs, showing Grimaldi and Oceanus Procellarum from a similar viewpoint to that of Figure 2.2, but at a lower sin angle and higher resolution. LO4-187 M (left) and LO4-181M (right), NASA, LPI, Image processing by the author.

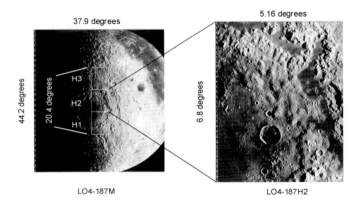

Figure 2.4. This figure shows the arrangement of the field of views of the frames taken by the Lunar Orbiter medium-resolution and high-resolution cameras. On the ground, the high-resolution frames were reconstructed in three subframes for easier handling by interpreters. The spacecraft was usually oriented so that the top of each frame or subframe pointed approximately to either the north or the south. By convention, the reconstructed frames are always oriented so that north is up.

The preexposed calibration strip, shown on the left here, is on the right if the H1 frame is to the north.

Phil Stooke of the University of Western Ontario has assembled a mosaic of Lunar Orbiter photos that cover the entire far side (Figure 2.5). He has converted each of these photographs into a common projection, selected the best area of each photo, and harmonized the brightness and contrast to produce the mosaic. This is an unusual image because it repre-

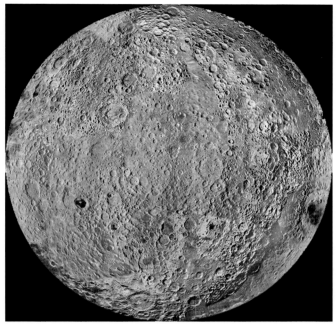

Figure 2.5. Mosaic of the far side of the Moon constructed of Lunar Orbiter photographs from all five missions. Lambert equal area projection. Stooke.

sents the spherical Moon as if it was illuminated at a constant sun angle everywhere. It demonstrates the full coverage of the Lunar Orbiter missions, at their various orbital altitudes. Of course, the resolution varies from point to point.

In the photography of the near side, the high-resolution images were contiguous for selected targets, with the medium-resolution images providing context. On Lunar Orbiter 4, the mission accomplished contiguous high-resolution coverage for the entire near side of the Moon. The far side photos were arranged to give contiguous coverage with medium-resolution photos from all five missions. Therefore, the high-resolution photos provided only sampling on the far side and only occasionally overlap. When they do overlap, this book often presents them as mosaics.

The entire process of transmitting imagery for Lunar Orbiter was analog. Only recently have these pictures been digitized. This was done by the Lunar Planetary Institute (LPI), using a digital camera on the large hard copy photographs archived at the Regional Planetary Image Facility (RPIF) there. The equivalent number of pixels across the width of the hard copy originals is about 5,000. The digitized images provide about 900 pixels in width. The precise count of pixels varies from frame to frame, since the pictures were taken with small margins and then cropped.

The process of scanning the negatives in the spacecraft and reconstructing them on the ground introduced artifacts that degrade the visual quality of the images (Figure 2.6).

Fortunately, the noise introduced by the artifacts is mostly in the north-south direction, while the signal produced by the slanting sunlight is mostly in the east-west direction. Therefore, it is possible to separate the signal from the noise and greatly increase the perceptual image quality (Figure 2.7). The method of doing this is described in Appendix A. All of the Lunar Orbiter photographs in this book and its companion book "Lunar Orbiter Atlas of the

Figure 2.6. Lunar Orbiter high-resolution subframe LO4-187H2 downloaded from the LPI web site, with the scanning and reconstruction artifacts.

Figure 2.7. After cleanup by the author, the signal showing the topography is intact, but the artifacts have been removed (see Appendix A for a description of the process).

Near Side of the Moon" have been cleaned in this manner. The cleaned images are all provided in the Compact Disc accompanying this book and their use for research and talks is encouraged.

Apollo

Apollo 16 took photos with three cameras, the mapping camera, the panoramic camera, and the hand-held Hasselblad. Figure 2.8 shows a mosaic of two successive exposures of the mapping camera. Apollo 15, 16, and 17 carried mapping cameras.

Clementine

Clementine's CCD camera system was very different from the Lunar Orbiters' film cameras. All the Clementine images in this book are mosaics constructed from the UV–vis (ultraviolet and visible range) camera. It took a field of view of $4.2°$ $× 5.6°$ with an array of $288 × 384$ pixels. The camera took a series of contiguous frames as the motion of the spacecraft moved in its polar orbit. The camera recorded overlapping images on the next orbit, as the Moon rotated beneath the spacecraft orbital plane. The scans were assembled into a mosaic after being transmitted to Earth and the resulting image was smoothed and corrected for the orbital geometry at USGS. The resolution varied from 100 to 500 m.

As mentioned above, the Clementine mission was designed for a high sun angle as the craft passed over the lunar equator, a desirable situation for the multispectral imagery that distinguishes mineral composition. However, the topography is washed out in this condition (Figure 2.9). In the zones

Figure 2.8. This is a mosaic of two Apollo 16 mapping camera frames covering the floor of Tsiolkovskiy. It clearly shows the contrast in albedo and texture between the rugged walls of the crater and its floor, flooded with very liquid lava.

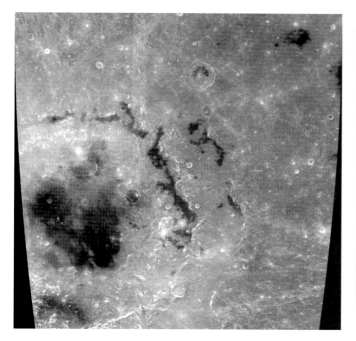

Figure 2.9. Clementine mosaic of the Orientale region, similar to the Lunar Orbiter image of Figure 2.3. USGS Map-a-Planet (sinusoidal projection, 750 nm).

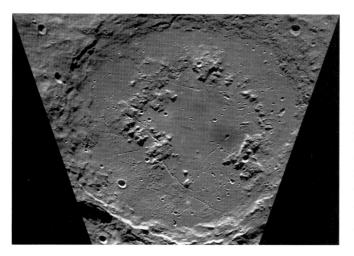

Figure 2.10. This Clementine mosaic shows the Schrödinger Basin. The approximate latitude is 75° so that the sun elevation is about 15°. The topography dominates the shading of this image. USGS Map-a-Planet (sinusoidal projection, 750 nm).

Figure 2.11. Photo of the far side by the Nozomi spacecraft. The ray patterns of Necco and King combine to form the bright splash below and to the left of center. Data courtesy of JAXA, ISIS, and the MIC Team. Processing courtesy of Ted Stryk.

near the north and south poles, beyond 55° in latitude from the equator, the sun elevation decreases from 35° to 0° as the pole is approached and so there is reasonable representation of the topography there (Figure 2.10).

Figure 2.9 shows a typical Clementine image in the equatorial area. The sun elevation is about 75°. Most of the shading is caused by albedo variations such as the dark shade of mare basalt contrasting with the light shade of highland anorthosite. Topography is washed out but rays of young craters are clearly shown. A particular characteristic of the Clementine imagery is that it is available at uniform resolution over the entire Moon, and is especially useful in organizing coverage of very large features like the South Pole-Aitken Basin.

Figure 2.10 shows a Clementine image near the South Pole, where the sun elevation is low and reveals the topography.

Nozomi

The unique picture of Figure 2.11 is the nearest thing to a "full Moon" picture of the far side.

Chapter 3

The History of the Moon and Its Features

This chapter is tutorial, to provide an understanding of the types of features found on the Moon and to introduce the special terminology of lunar geology that is used in the notes that accompany the photographs. As terms are introduced, they are shown in bold print. These terms are also defined in the glossary.

3.1. Origin of the Moon

After considering other possibilities, the consensus view is that the Moon began as a small rocky planet sometimes called Theia (the goddess) after Theia Euryphaessa, the mother of Selene (the goddess of the Moon) in Greek mythology. Originally in an orbit somewhere near what is now the asteroid belt, interaction with other bodies disrupted that orbit and turned Theia toward the Sun. On the way, it encountered Earth, striking a glancing blow that vaporized, melted, and pulverized Theia and part of Earth's crust. The Moon coalesced from that part of the resulting cloud that orbited Earth. Part of the evidence for this is the deficiency of volatile materials like water and sulfides in the Moon; they must have stayed in the cloud and were dissipated into space or back to Earth.

3.2. The Magma Ocean

As the Moon coalesced, it would have formed a very hot molten body, nearly spherical but slightly oblate because of its rotation, and perhaps distorted additionally by the proximity of Earth's gravity. The molten mixture of elements that existed in this early time, 4 billion years ago, is called the **Magma Ocean**. As the Magma Ocean cooled, the heavier compounds drifted toward the core and the lighter ones rose toward the surface. Some of the heaviest and most common elements like iron, perhaps in combination with some sulfur, settled to the core. Lighter elements, especially silicon and oxygen in the form of silicates of metals like aluminum, started to form a crust that established a discontinuity in density like that between the crust of Earth and its mantle. On Earth, that boundary is called the Mohorovičić discontinuity.

Our current understanding of the very early formation of the Moon does not explain the differences between its two sides. The difference must have come about later.

3.3. Bombardment

Early in the history of the inner solar system, there was an extended period of heavy bombardment by bodies smaller than planets, probably fragments of bodies that were of planetary size or perhaps smaller that collided and fragmented. The asteroid belt may be a remainder of that period, composed of small bodies that achieved a relatively stable orbit. A significant fraction of that material was thrown toward the sun, impacting all of the rocky planets and their moons. This process is ongoing today, thankfully at a lesser rate but occasionally penetrating roofs and denting cars, although more commonly splashing in the ocean. Such a body approaches a planet or moon with at least the escape velocity of the planet, the velocity that accounts for the potential energy lost as a body entered the planet's gravity field. In the case of the Earth and Moon, typical impact velocities are about $20\,km\,s^{-1}$ (12.5 miles per second), far greater than the speed of sound in solid material. The resulting collision is called a **hypervelocity impact**, causing a crater to be formed in the **target surface** that was in the way of the speeding **impactor**.

3.4. The Near Side Megabasin

I propose that the Near Side Megabasin, an impact basin whose central depression covers more than half of the Moon, has differentiated the two sides of the Moon. It could have caused the depression on the near side (where the large maria occur) and also the thicker, mounded crust of the far side. The shape of the far side was, of course, further modified by the South Pole-Aitken Basin to establish the current configuration. The general idea is not new (Wood, 1973; Cadogan, 1974; Whitaker, 1981), but the Near Side Megabasin is larger than former proposals of the Gargantuan Basin and the Procellarum Basin and is of the appropriate scale to do the job. Oceanus Procellarum is entirely within this newly proposed basin, formed by the evolution of the internal depression. The rim of the depression is nearly all on the far side. The center of impact is not quite centered on the near side of the Moon but is at 8.5° north latitude and 22° east longitude. More detailed parameters, along with the analysis leading to this hypothesis, are described in Chapter 13. In that chapter, a model of impact basins is presented and accepted scaling laws are used to show that the large-scale shape of the present Moon is consistent with the **ejecta blanket** that such a large basin would produce. The mound of **ejecta** on the far side conforms quite well with the shape of this model.

The term **megabasin** is used in different technical contexts to mean a depression in terrain (or a minimum of a mathematical function) that is so large that it contains smaller basins within it. In this sense, the Moon contains another megabasin, the South Pole-Aitken Basin, the result of another giant **impactor** striking the Moon. This impact occurred later than the earlier and larger one that is proposed to have produced the Near Side Megabasin.

3.5. The South Pole-Aitken Basin

The South Pole-Aitken Basin extends for about 80° across the far side Southern Hemisphere. Its floor is largely exposed despite shallow deposits of mare in the floors of the deeper basins and craters within it.

3.6. Impact Features: Craters and Basins

Craters produced by hypervelocity impacts all develop according to similar laws. Indeed, similar craters are observed from any explosive release of energy such as underground tests of nuclear bombs or explosive release of energy from volcanoes. The shapes of impact features are nearly independent of the nature of their impactors and nearly independent of the approach angle, if that angle is more than 25° from the horizontal. Some differences due to impactor composition are introduced by entry through an atmosphere, but that is not relevant to the Moon.

There are three primary divisions of the features of an impact feature: the transient crater, the apparent crater, and the ejecta.

The **transient crater** forms quickly, within seconds or minutes, depending on size. It is roughly hemispherical below the surface and produces a turbulent zone of fractured, melted, and pulverized material in the target. Its diameter relative to that of the impactor depends on the gravity of the planet, but for Earth, estimates range from 5 to 10 times the diameter of the impactor. It follows, of course, that the impactor contributes less than 1% of the material that remains in the transient crater.

The **apparent crater** is the cavity that is in the visible surface after the event. It results from the material that is removed when the expanding energy released by the impact reaches the surface and, not meeting further resistance, ejects a layer of the material there. The apparent crater is much shallower than the transient crater for large craters and basins.

Ejecta is the material thrown out from the apparent crater. Some of this material forms a rim around the crater whose depth is about 20% of the depth of the apparent crater,

measured from the original target surface. Ejected material that falls beyond this rim, out to about twice the rim radius from the center of the feature, forms a continuous deposit called the ejecta blanket. The depth of ejecta falls off rapidly beyond the rim. This ejecta blanket often has ridges and troughs radial to the center of the apparent crater. Beyond the ejecta blanket, the ejected debris consists of powdery material that may produce **plains** and **rays.** Boulders that are ejected produce, upon landing, their own s**econdary craters.** These "secondaries" sometimes land in radial strings or chains, called **catena.**

For a given type of target surface, crater morphology undergoes a progression from a simple symmetrical cavity to a cavity with a central peak, and then to a disrupted surface that forms an internal ring. Beyond a certain size, multiple rings appear both inside and outside of the apparent depression. Such impact features are called **ringed basins.** Lunar impact features larger than 300 km in diameter usually exhibit rings (or arcs of rings) and those smaller than 300 km usually do not. Therefore, the term **basin** is used for features with diameters larger than 300 m and the term **crater** is used for features that are smaller.

Throughout this book, attention is drawn to the examples of these structures and how the structures of neighboring impact features interact. The reader is encouraged to form personal views of the patterns of these structures.

3.7. Mare Basalt

Some deep craters and basins on the far side, as well as most such features on the near side, are filled with **mare** material, lava extruded from below. As we have mentioned, there are fewer and smaller areas of mare on the far side, relative to that on the near side. The largest, in the central floor of the South Pole-Aitken Basin, is about the size of Mare Imbrium (1,023 km in diameter). The Imbrium Basin is filled all the way to its rim but the larger South Pole-Atken Basin just has concentrations of mare in the deeper craters and basins of its central floor. All other areas of mare on the far side are much smaller; the largest of them is Mare Moscoviense, 277 km in diameter.

Mare material is basaltic, like our ocean floor. It originates from the heavier material in the **mantle,** which is richer in iron and magnesium and lacking in aluminum and silicates, in relation to the crust. As the Magma Ocean cooled, the mantle material settled because of its higher density. However, radioactive elements of the uranium series liberated heat energy, remelting parts of the mantle. The molten material expanded, became less dense, and rose to the surface, where it cooled. The radioactive elements leave their signature: high concentrations of the products of radioactive decay of uranium, like thorium. A related set of elements is known as **KREEP,** for potassium (chemical symbol K), rare earth elements (REE), and phosphorus (chemical symbol P). They tend to stick together, and are enriched when mantle rock is remelted, because they do not crystallize easily but they do melt easily. These two types of tracers, both detected in the maria, are telltales of the history of basalt that has come to the surface. The KREEP tells us that the material was remelted and the thorium tells us what remelted it.

The high concentrations of thorium in the mare material of the South Pole-Aitken Basin tells us that uranium, normally low in the mantle, has been stirred up to be nearer the surface by the large transient crater of that basin. The same sort of turbulent lifting of heavy elements probably occurred beneath the floor of the Near Side Megabasin, which has an even stronger thorium and KREEP signature on its surface.

3.8. The Ages of the Lunar Features

An interesting attribute of a lunar feature is the time it was formed. The ages of different features on the Moon are inferred from several sources.

The most precise ages are determined from measurement of isotope ratios in our precious rock samples; association of the samples with specific features establishes their time of formation. Since no samples have been returned from the far side, estimates of the age of events there can only be determined by estimates of ages relative to features of the near side.

When rock samples are unavailable, geologists have resort to less precise measures. For example, sharply defined features are recent because meteorite bombardment softens the edges of features by a process called **mass-wasting**. And counts of crater densities are used to estimate the time a surface has been exposed to bombardment.

A very powerful principle used by geologists is **superposition**. Where one feature overlays another, it is considered to be a younger feature, unless there is evidence to the contrary. There are exceptions. For example, an impact may throw a boulder from an older layer of rock onto a younger layer. But the general rule is useful, especially in the emplacement of the ejecta blanket of one basin on top of the ejecta layer of another. The vertical sequence of layers from different sources is called **stratigraphy**.

The principle of superposition has other implications than aging. When one layer is deposited on another, the topography of the surface reflects the configuration of the lower layer, as modified by variations in depth of the upper layer. For example, it is sometimes possible to detect the shape of a crater that has been buried by an ejecta blanket. If a crater forms on a slope or rim of an earlier basin, its shape is superimposed on the earlier shape.

Sometimes one event causes another, even though superposition is not involved. For example, the shock wave of a younger impact event can be seen to have caused a landslide at the rim of older nearby crater.

The notes accompanying the photos in this book frequently point out examples of superposition and causality.

Named Age Ranges of the Moon

Systematic study of stratigraphy, together with other clues, establishes a chain of evidence that leads to a relative age range of most of the features of the Moon. These age ranges, called periods, are named for archetype features. In order from the oldest to the younger, they are the **pre-Nectarian Period**, **Nectarian Period**, **Early Imbrian Period**, **Late Imbrian Period**, **Eratosthenian Period**, and **Copernican Period**.

The impact of the Nectarian Basin defines the boundary between the pre-Nectarian and Nectarian Periods. The Imbrium Basin event introduces the Early Imbrian Period and the Orientale Basin event (the youngest basin) begins the Late Imbrian Period. The Eratosthenes crater is the archetype of the Eratosthenian Period and Copernicus is the archetype of the Copernican Period. The Early and Late Imbrian Periods are often called epochs.

The far-flung ejecta of these giant basins overlaps and provides a precise means for judging relative ages. A summary of estimates for the absolute age of the periods of lunar history is given in Table 3.1 (Wilhelms, 1987). Of course, these absolute ages have uncertainty ranges, but these are not shown in the table.

After the Orientale event 3.8 billion years ago, there were no further basins formed; in fact, all subsequent craters are smaller than 190 km in diameter. The time of heavy bombardment had passed. The ejecta blankets of the smaller craters formed in later times no longer overlapped, except rarely. Relative ages of the younger features are established by judgements about the degree of erosion of them; they become less sharp as they are hit by the continuing arrival of small impactors. If a feature has a similar degree of erosion to the crater Eratosthenes, it is judged to be of the Eratosthenian Period. If it appears to be as fresh or fresher than the crater Copernicus, it is assigned to the Copernican Period.

The designations of these age periods predate the views of the far side of the Moon and so they are all defined in reference to features on the near side. The relative ages of far side features are based on the stratigraphic relations, feature by feature, from one basin's ejecta to another across the limbs all the way to the far side. For this purpose, especially important basins are Crisium, Orientale, and Imbrium, since their deposits fall on the far side. Features on the far side that are younger than the Orientale Basin are assigned to the Late Imbrian, Eratosthenian, or Copernican Periods depending on the sharpness of their features (mass-wasting is a form of erosion that smoothes features with time) and on counts of subsequent (superposed) craters.

A somewhat controversial indication of age among younger impact features (late in the Eratosthenian Period or early in the Copernican Period) is the presence of rays of bright material forming a star-shaped pattern. One hypothesis is that these rays dim uniformly with age and their presence or absence can be used to establish relative age. This rule has met with contradiction; some craters that appear fresher than others have lesser rays.

Period	Age range (billions of years ago)	Interval (millions of years)
Pre-Nectarian	4.6 to 3.92	680
Nectarian	3.92 to 3.85	70
Early Imbrian	3.85 to 3.80	50
Late Imbrian	3.80 to 3.15	650
Eratosthenian	3.15 to ~1.0	~2,150
Copernican	~1.0 to present	~1,000

Table 3.1. Estimates of absolute ages of the periods.

Recent research (Hawke, 2007) suggests that there are two mechanisms for rays to darken and fade. All rays start to darken with time because exposure of the fresh powdered material to the solar wind causes the iron in the grains to aggregate, absorbing radiation. In time, the effect stabilizes, reaching a condition called maturity. The second effect relates to impacts in areas composed of anorthositic highlands near maria. The anorthositic minerals are inherently lighter than the basaltic mare, even after they become mature. Consequently, they only dim as meteoroid bombardment mixes the rays with the substrate, a process that takes much longer than the solar wind effect. These rays are called compositional rays. The time required to thoroughly mix a compositional ray with its substrate so that it fades depends on the thickness of the ray, and therefore on the size of the associated impact feature.

The suggestion has been made that the duration of the Copernican Period be redefined in light of the new understanding. It is proposed (Hawke, 2007) that the start of the Copernican Period be defined to be the time of the impact event of Copernicus, whose rays are now nearly mature but have bright compositional contrast. In the past, Copernicus has served as the archetype crater of the Copernican Period, which was allowed to extend prior to the Copernicus impact event. It will be interesting to see how this proposal is adopted in future geologic research and mapping work. If it is, then craters that are located on the far side, provided they are outside of the South Pole-Aitken Basin and the very small other areas of mare, will be considered Copernican only if they have visible (immature) rays. This is because the composition of those areas of the far side is very uniform, so that the appearance of compositional rays blends into that of the substrate.

Since the discovery of the South Pole-Aitken Basin and the proposal of the Near Side Megabasin, it might be useful to designate new periods to decompose the pre-Nectarian Period. For example, one could define a period between the first formation of the lunar crust to the Near Side Megabasin event, a period between that event and the South Pole-Aitken Basin event, and a period between that event and the Nectaris Basin event.

Chapter 4

Regions of the Far Side

The photographs presented in this book are grouped into regions, as shown in Figure 4.1. Each region is covered in one chapter. In the order of their chapters, they are the following.

4.1. The Western Far Side Region

This chapter includes photographs of the Earth taken from space, as well as the basins along the limb. The large craters Tsiolkovskiy and Gagarin are in this region, as is the Mendeleev Basin.

4.2. The Korolev Basin Region

The Korolev Basin is at the top of the 5 km bulge on the far side of the Moon.

4.3. The South Pole-Aitken Basin and the South Polar Region

The South Pole-Aitken Basin is enormous, stretching from nearly the equator down to the South Pole and beyond. It contains the largest concentrations of mare on the far side. The South Polar Region includes craters whose floors are permanently shadowed and show a high concentration of hydrogen. Some of this hydrogen may be in the form of water ice crystals just below the surface.

4.4. The Northwestern Far Side Region

This region is interesting for the interplay of the Humboldtianum Basin with the mare-filled Moscoviense Basin.

4.5. The Eastern Far Side Region

This region shows the basins such as Birkoff and Hertzsprung, whose interleaving ejecta blankets have been preserved, without resurfacing by mare flows as has obscured the floors of the near side basins.

4.6. The North Polar Region

There are several basins along its southern edge. The pole itself is within the ejecta blanket of the Imbrium Basin of the northern near side.

4.7. The Orientale Limb Region

This region, near the West Limb (as we see it from Earth), is dominated by the spectacular Orientale Basin, the last and best preserved of the large basins.

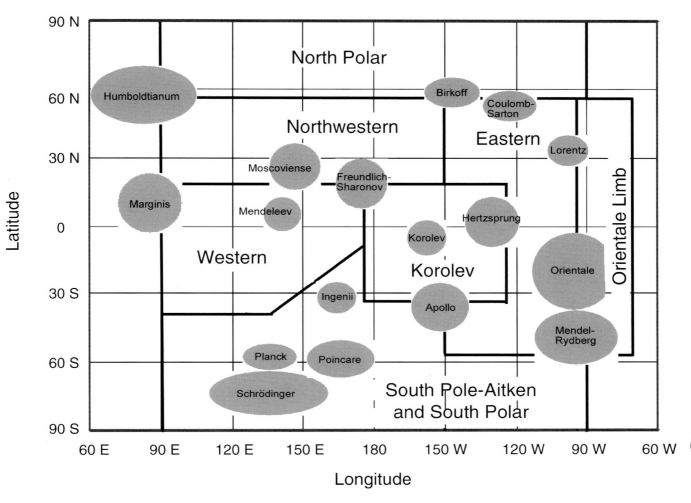

Figure 4.1. The regions covered in each chapter are shown above, with many of the basins shown. The coverage areas of each chapter overlap. Specific coverage of each region is described in more detail early in each chapter.

Chapter 5

Nomenclature and Conventions of This Book

A number of conventions have been adopted to show the lunar features that appear in these images.

5.1. Identification of Named Features

In general, each chapter covers its region with two sets of images, those intended for overlapping, contiguous coverage, and those that are samples at higher resolution. Each contiguous coverage image is presented twice, once with annotations and once unmarked.

Annotated Images

The annotated images show the named features recognized by the International Astronomical Union (IAU). Each feature is outlined, either by the staff of the Lunar and Planetary Institute or by the present author. A few small named features (less than 50 km in diameter) are not annotated, for practical reasons. Since the images overlap, some features appear on more than one image. All annotated features are listed in an index.

Spelling of Feature Names

The spelling of the feature names is that of the IAU, as it appears in the catalog maintained by the United States Geological Survey (USGS). The text follows the conventions of the annotations. In some cases, the spelling of IAU names differs from the spelling used in the text, when persons the features are named for are mentioned. An example of this distinction is that one far side crater is called Tsiolkovskiy, while the text refers to Konstantin E. Tsiolkovsky, the Russian rocket scientist. The spelling is the same in Russian; the difference lies in diverse transliteration rules used by the IAU and by translators to the English language.

Features Identified on High-Resolution Images

Only one copy of most of the high-resolution samples is provided. To identify features that are referenced in the notes with minimal obscuration of the images, they are marked by short letter codes that are indicated both in the images and after the feature name in the text. Capital letters are used for IAU named features and small letters are used for other features mentioned in the notes.

Lettered Craters

Smaller craters of interest to researchers but too numerous to be assigned specific names are designated by the name of a nearby feature, followed by a single letter (Tsiolkovskiy W, for example). The IAU recognizes these designations. The lettered craters are not annotated in this book unless they are specifically mentioned in a note.

On the near side of the Moon, these letters have not been assigned systematically. However, on the far side, a convention proposed by Ewen Whitaker (Whitaker, 1999) has been adopted. In this convention, the letter is chosen according to the approximate direction from the named feature, using the 15° divisions of a 24-h clock face as a guide. The numbers 1 through 24 are replaced by the letters A through Z, with the omission of I and O. The direction indicated by Z points due north.

5.2. Identification of Basins

The geologists of the USGS Astrogeology Branch maintain the names of basins. The IAU does not formally identify basins, partly because there is no universally accepted way to distinguish small basins from large craters. Where a basin coincides with a named feature or surrounds a named mare, that name is adopted for the basin. Where it is not, the practice is to name the basin for two named craters near its periphery (the Keeler-Heaviside Basin, for example). On annotated images, basins with compound names are usually shown as dashed outlines.

The rim of the Near Side Megabasin, as proposed by the present author, is indicated by dashed lines, in the same manner as the other basins.

5.3. Latitude and Longitude

The latitude and longitude of each annotated image is indicated either by edge notes or by overlaid meridians and parallels.

The high-resolution images are not so marked, but their locations are clearly identified in the relevant medium-resolution images, as described in the introduction to each chapter.

5.4. Scale and Age

Because so many of the images in this book are taken from high orbits and are often taken at oblique angles, scale changes widely within most of the images. A scale bar would be misleading. Consequently, care has been taken to identify the diameter of a few features within each image.

The diameter and age of many features are given in a "key" table that accompanies the notes of each annotated image. The diameter is usually taken from the IAU catalog. The diameter of basins is as given by the USGS (Wilhelms, 1987). In some cases, these diameters differ slightly from the IAU diameters when a basin coincides with a named feature, but the USGS diameter is always used for basins, for consistency.

The age period assigned to features is also taken from the USGS (color plates of Wilhelms, 1987). An attempt has been made to include all features named in these plates in the keys, as well as some features identified, but not named, on the plates.

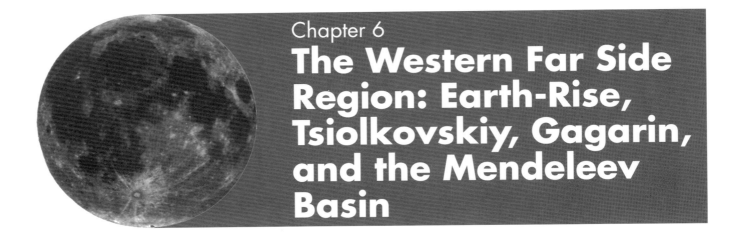

The Western Far Side Region: Earth-Rise, Tsiolkovskiy, Gagarin, and the Mendeleev Basin

The Far Side **18** of the Moon

6.1. Overview

This region is west of the center of the far side, if you are looking toward Earth (as the Apollo astronauts were) but of course it is out of sight beyond the East Limb if you are on Earth looking toward the Moon. The region is the most heavily photographed part of the far side because it was sunlit when the Lunar Orbiters and the Apollo Command Modules passed over it in their near-equatorial orbits.

Most of this region is shown in Figure 6.1. The Mendeleev Basin near the top dominates the view. The crater Tsiolkovskiy at the lower left contains one of the largest areas of dark, relatively fresh, mare pierced by its central peak. The crater is named after Konstantin Tsiolkovsky. "Tsiolkovsky" is the most common English transliteration of his name, but the International Astronomical Union, using another rule, designates the crater's name as Tsiolkovskiy. Tsiolkovsky was a Polish-Russian rocket and space flight theoretician in the early twentieth century. The crater Gagarin at the lower right has a smaller crater, Isaev, within its northwest sector. This crater has penetrated the floor of Gagarin and has produced a crustal weakness that permitted lava to penetrate and form a dark mare surface.

Like nearly all of the far side, the area is completely covered with impact craters; in fact, they obscure each other so that only the younger craters can be seen clearly. The older basin, Mendeleev, near the top of Figure 6.1 has probably had its floor flooded with mare, but then ejecta from younger basins has covered the dark mare with lighter material. This hidden feature is sometimes called cryptomare.

The rim of the Near Side Megabasin runs through this region, at about 120° east longitude, a little west of Tsiolkovskiy and about a diameter west of Mendeleev. This rim, a rise of about a kilometer, runs along the sudden transition between the bright sunlit washed out area and the gray area to the right of the rim.

6.2. Photos

Lunar Orbiter Medium-Resolution Frames

The approximate coverage of the medium-resolution photos targeted for the Western Far Side region is shown in Figure 6.2. The actual medium-resolution photos are shown starting on page 21. All of the medium-resolution photos are shown before the corresponding high-resolution frames, to avoid breaking the continuity of contiguous coverage and annotation of the medium-resolution photos.

A key table is provided to identify the dimensions and age of prominent features in each annotated picture.

Lunar Orbiter High-Resolution Subframes

There are three high-resolution subframes for each of the medium-resolution Lunar Orbiter frames that are labeled in Figure 6.2, and three such subframes for LO1-117M, whose principal ground point is off the chart. In each case, the high-resolution frame H2 is centered on the principal ground point, H1 is to the north of H2, and H3 is to the south of the central frame. The three subframes of LO1-102H and LO1-117H are reassembled into mosaics because they show the especially interesting photos of the Earth rising over the lunar horizon.

The high-resolution photos are marked with code letters to identify features that are mentioned in the notes that accompany each picture. At least one dimensioned feature is mentioned in each note in order to establish scale for the photo.

Figure 6.1. This is a photo-mosaic of two Lunar Orbiter medium-resolution frames. It shows much of the western far side region from 90° east longitude to nearly 180° east longitude (the terminator is at 164° east longitude). The top of the picture is at 16° north latitude and the bottom is at 26° south latitude. The crescent Earth is in the background (LO1-116M and LO1-117M, NASA, LPI).

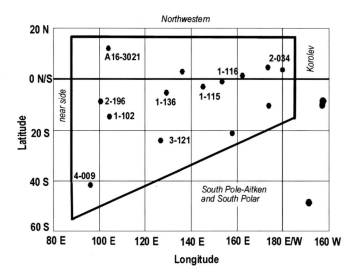

Figure 6.2. This figure labels the principal ground points and approximate total area of coverage of the medium-resolution Lunar Orbiter photos included in this chapter. LO1-102M and LO1-117M show Earth, seen from behind the Moon (the principal ground point of LO1-117M is off the Moon). A16-3021, taken by the Apollo 16 mapping camera, LO2-196, and LO4-009 cover the western limb of the far side. The unlabeled points near the equator relate to Lunar Orbiter mission 4 photos that were too low in resolution to contribute to this chapter. The two unlabeled points near 160° east longitude correspond to photos in the South Pole-Aitken region (Chapter 8).

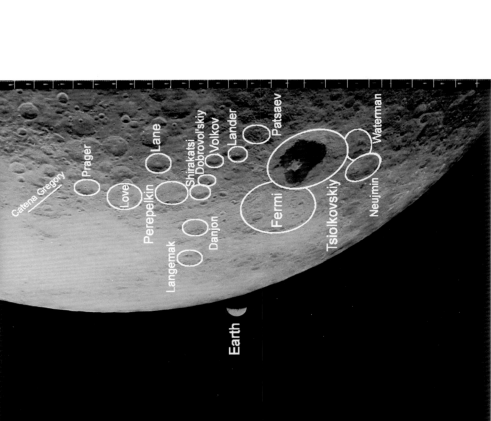

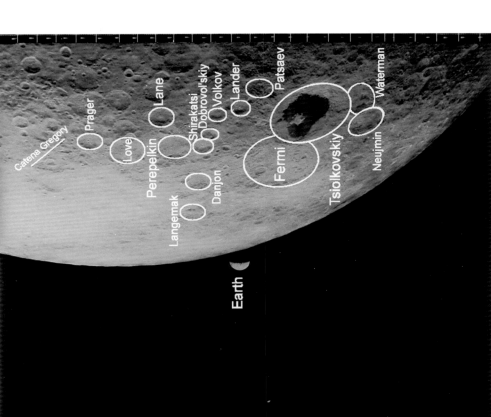

LO1-102M   Sun Elevation: 68.68°   Distance: 2107.55 km

Lunar Orbiter 1 took the first photo of Earth from behind the Moon on August 22, 1966. High-resolution frame 1-102H2 shows detail of Earth's cloud cover. In the foreground of this medium-resolution photo are dramatic oblique views of Tsiolkovskiy (flooded with mare), its neighbor Fermi, and the chain of craters Catena Gregory.

Key:

Pre-Nectarian
Fermi, 183 km

Late Imbrian
Lane, 55 km
Langemak, 97 km
Tsiolkovskiy, 185 km

Eratosthenian
Shirakatsi, 51 km

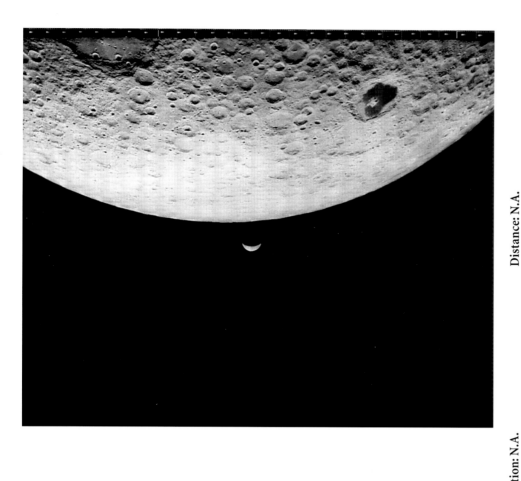

LOI-117M Sun Elevation: N.A. Distance: N.A.

Lunar Orbiter 1 took this photo a little later in the mission, from a point further north. The principle point (center) of the photo is on Earth, at a distance of about 240,000 km. The Mendeleev Basin has joined the features in the foreground.

Key:

Pre-Nectarian
Fermi, 183 km

Nectarian
Mendeleev Basin, 330 km

Late Imbrian
Lane, 55 km
Langemak, 97 km
Tsiolkovskiy, 185 km

Eratosthenian
Shirakatsi, 51 km

Copernican
Hartmann, 76 km
King, 76 km

The Western 21 Far Side Region

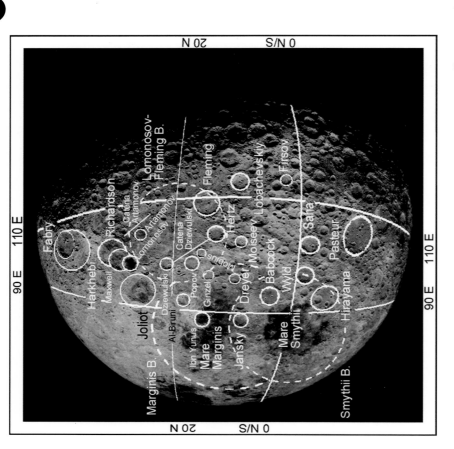

Apollo 16 mapping camera 3021

This photo from Apollo 16 shows the eastern limb of the Moon north of the equator. Mare flowed into the lower parts of the Smythii and Marginis Basins in the Late Imbrian and Eratosthenian Periods. The boundary where the level crust that fills the lower cavity of the Near Side Megabasin meets the slope that rises toward the rim to the east is approximately along the 110°E meridian here. Mare lava filled the floors of Al Biruni, Artamonov, Joliot, Lomonosov, Maxwell, and Richardson in the Late Imbrian Period.

Key:

Pre-Nectarian
Fabry, № 184 km
Harkhebi, 237 km
Hirayama, 132 km
Joliot, № 164 km
Lomonosov-Fleming
    Basin, 620 km
Marginis Basin, 580 km
Pasteur, 224 km
Richardson, № 141 km
Smythii Basin, 840 km

Nectarian
Fleming, № 106 km
Maxwell, 107 km

Early Imbrian
Firsov, 51 km
Lomonosov, 92 km

Late Imbrian
Mare Marginis
Mare Smythii

Eratosthenian
Mare Marginis (later flows)
Mare Smythii (later flows)
Moiseev, 59 km

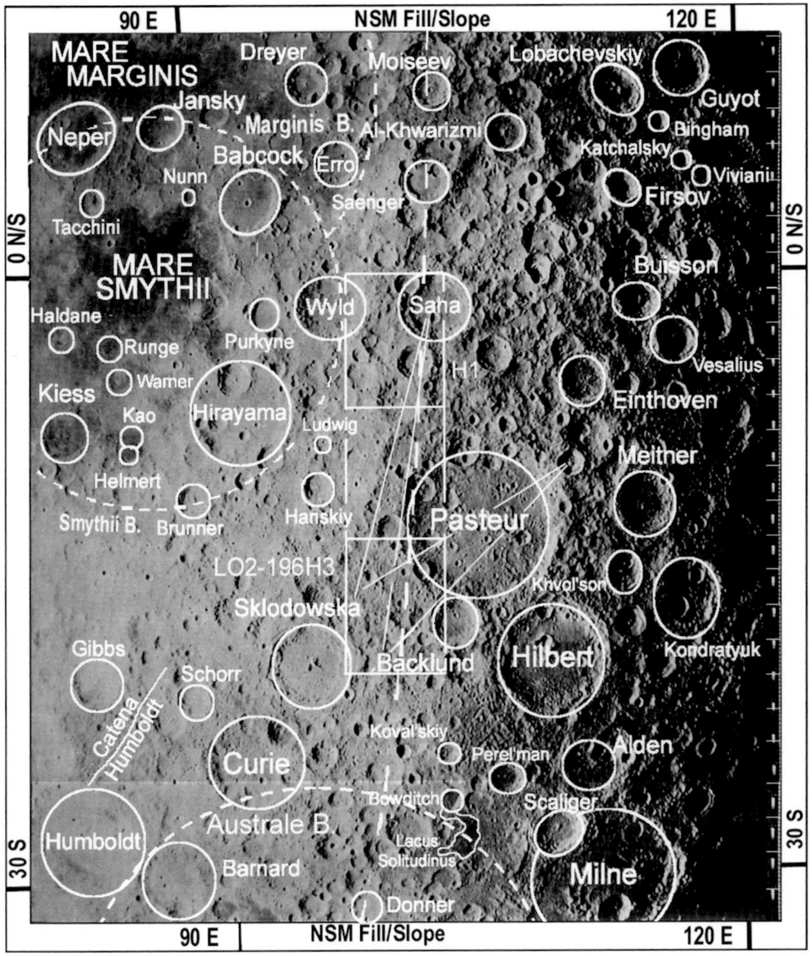

Distance: 1519.06 km

Sun Elevation: 19.86°

LO2-196M

The eastern limb of the Moon (90° east longitude) is, of course, the western limb of the far side. The eastern rims of the overlapping Marginis and Smythii Basins can be seen here. The Smythii impact came later, as evidenced by its clearer rim, but parts of both rims are visible (an example of the principle of superposition). The vertical dashed line marks the transition between the level fill of the proposed Near Side Megabasin and the unfilled slope of the basin rising toward its rim. The rectangles outline the high-resolution subframes H1, H2, and H3 (from the top).

The lines from Saha and Pasteur D, the crater northeast of Pasteur, show fields of secondary craters that can be seen in high-resolution subframe LO2-196H3.

Key to far side features beyond 90°

Pre-Nectarian
Curie, 151 km
Hirayama, 132 km
Milne, 272 km
Pasteur, 224 km

Nectarian
Al-Khwarizmi, 65 km
Meitner, 87 km
Saha, 99 km
Vesalius, 61 km
Hilbert, 151 km

Early Imbrian
Alden, 104 km
Brunner, 53 km

Late Imbrian
Scaliger, 84 km
Sklowdowska, 127 km
Lacus Solitudinis

LO4-009M

Sun Elevation: 17.81°

Distance: 3011.79 km

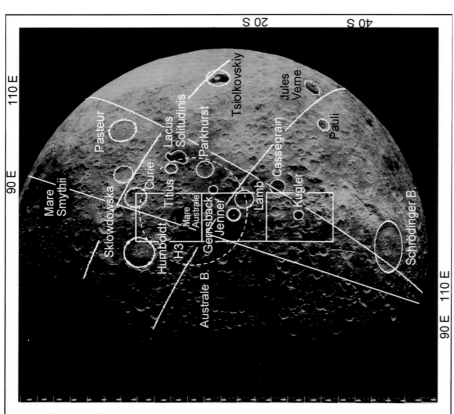

(Labels on map: 90 E, 110 E, Mare Smythii, Pasteur, Sklowdowska, Curie, Titius, Lacus Solitudinis, Humboldt, H3, Mare Australe, Australe B., Gernsback, Jenner, Parkhurst, Lamb, Cassegrain, Kugler, Tsiolkovskiy, Jules Verne, Pauli, Schrödinger B., 20 S, 40 S, 90 E 110 E)

Key:

Pre-Nectarian
Australe Basin, 880 km
Curie, 150 km
Milne, 272 km
Pasteur, 224 km

Nectarian
Kugler, 65 km

Late Imbrian
Humboldt, 189 km
Jenner, 71 km
Lacus Solitudinis
Mare Australe
Scklowdowska, 127 km

This photo completes the coverage of the eastern limb in the Western Far Side Region, down to 40° south latitude. The Australe Basin dominates this area, a very old basin that has been partially but not completely flooded with mare. East of Australe, like the other basins to the north, mare is rare, appearing only in a few deep craters like Tsiolkovskiy, until the very large South Pole-Aitken Basin is reached beyond Jules Verne.

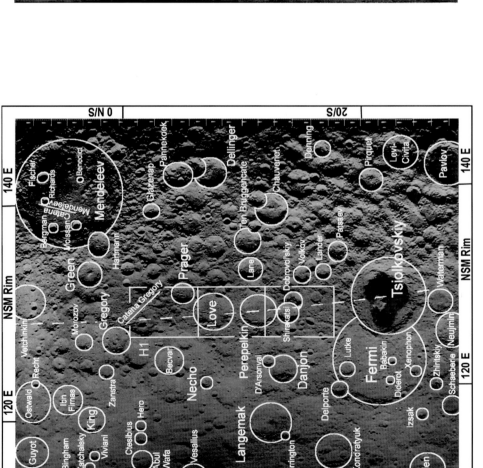

Distance: 1373.43 km

Sun Elevation: 20.65°　　LO1-136M

The rim of the proposed Near Side Megabasin rises to the east of the dashed line. Fermi has formed in the slope of the Near Side Megabasin. Subsequently, Tsiolkovskiy impacted the rim of Fermi and the much higher rim of the giant basin; as a result, the eastern rim of Tsiolkovskiy is higher than its western rim. Tsiolkovky and Pavlov are deep because they struck porous ejecta; as a result, they have been flooded with lava.

Distance: 1533.85 km

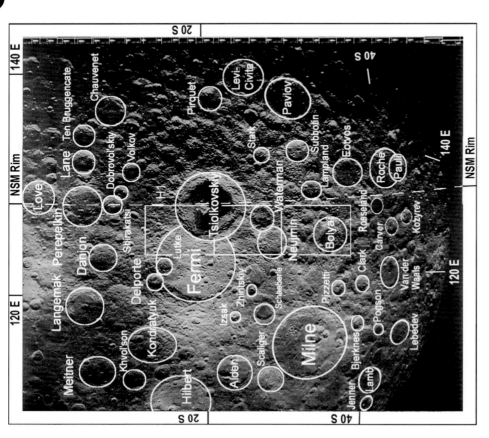

LO3-121M

Sun Elevation: 19.37°

Key:

Pre-Nectarian
Fermi, 183 km
Milne, 183 km

Nectarian
Pavlov, 148 km

Early Imbrian
Pauli, 84 km
Roche, 160 km

Late Imbrian
Bjerknes, 48 km
Pauli mare
Tsiolkovskiy, 185 km
Tsiolkovskiy mare

Eratosthenian
Izsak, 30 km
Shirakatsi, 51 km

See LO3-121H1 for detail of the floor, rim, and ejecta blanket of Tsiolkovskiy. Relative ages of many large craters in this region have been estimated by their shapes and how their ejecta falls on the others. Milne came early in the pre-Nectarian Period. It is a large crater whose inner ring illustrates the transition between crater and basin. Pavlov has been assigned to the next period, the Nectarian. Then came Roche and Pauli, then Scallinger and Tsiolkovskiy. Later, in the Late Imbrian Period, mare flooded the floors of Tsiokovsky, Isaev, and Pauli.

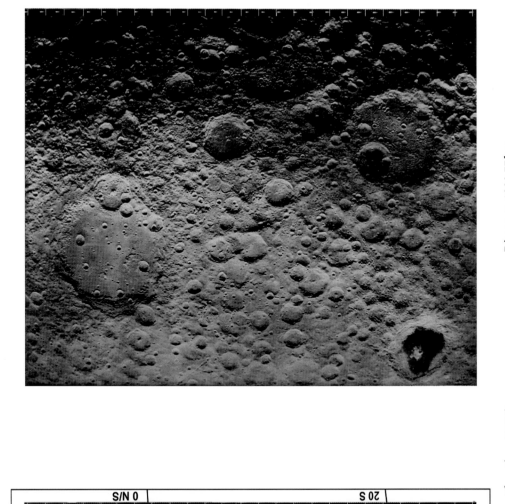

Distance: 1398.77 km

Sun Elevation: 18.96°

LOI-115M

Gagarin is a very old (pre-Nectarian) crater that has been more recently struck by Isaev, which has further penetrated its floor and allowed a flow of lava to enter from below.

Key:

Pre-Nectarian
Gagarin, 265 km
Keeler-Heaviside Basin, 780 km

Nectarian
Chaplygin, 137 km
Isaeve, 90 km
Mendeleev Basin, 330 km

Early Imbrian
Keeler, 160 km
Marconi, 66 km

Late Imbrian
Isaeve mare
Tsiolkovskiy, 185 km

Eratosthenian
Shirakatsi, 51 km

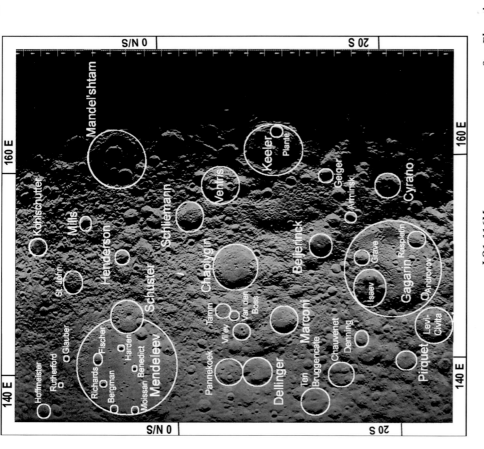

Distance: 1492.62km

Sun Elevation: 10.86°

LO1-116M

Key:

Pre-Nectarian
Cyrano, 80km
Gagarin, 265km
Ventris, 95km

Nectarian
Chaplygin, 137km
Mendeleev Basin, 330km

Early Imbrian
Keeler, 160km
Marconi, 66km

Eratosthenian
Izsak, 30km

The chain of craters between Gagarin and Marconi might be due to boulders from the Ingenii Basin to the southeast. Keeler has added its ejecta to the rim and floor of Ventris. Chaplygin is intermediate in age. It still shows signs of terracing, but those patterns are nearly eroded away. The Mendeleev Basin has also formed in an area of deep ejecta and been flooded with lava, but its surface is lighter, because it now has a thin layer of ejecta from other craters or basins.

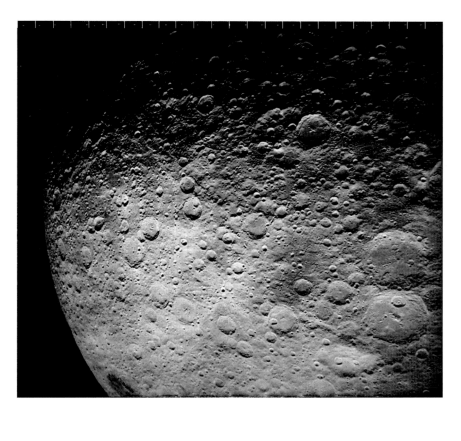

Distance: 1570.57 km

LO2-034M   Sun Elevation: 19.20°

The trough northeast of Ventris (also visible in LO1-116M) is radial to Gagarin. Striations across the upper half of this photo, running between the upper left corner and Krasovskiy, Lipskiy, and Daedalus, come partly from the Moscoviense Basin and partly from the Korolev Basin (beyond Icarus to the southeast). The dark area around Dewar, north of Keeler and Heaviside, is both low in albedo and depressed in elevation. Mare basalt may have invaded the area.

The differences in brightness and contrast in these two dramatic pictures of Earth are due to the way in which the Moon absorbs light (the lunar photometric function). Because the lunar horizon is nearly aligned with the direction of illumination in photo LO1-117, there is a "hot spot" there. The brightness level of the print has been reduced to capture as much detail as possible, making the Earth seem dimmer, although, of course, the actual brightness of the cloud layer was the same in the two photos.

The cloud pattern has also changed in the interval between pictures, probably due more to the rotation of the Earth than to changes in the weather patterns.

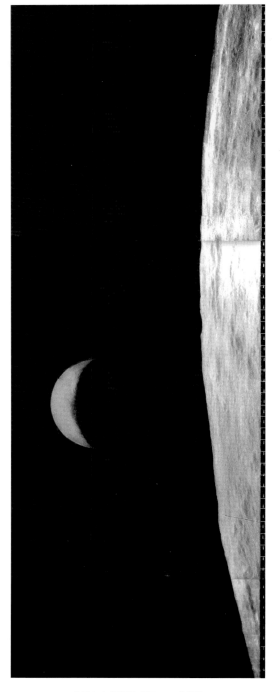

LO1-102H1, H2, and H3

LO1-117H1, H2, and H3

LO1-102H2   Sun Elevation: 68.68°   Distance: 2107.55 km

This spectacular view of Earth was the first view from behind the Moon. Later, Apollo took color pictures of this same view, giving rise to the characterization of Earth as a "Big blue marble."

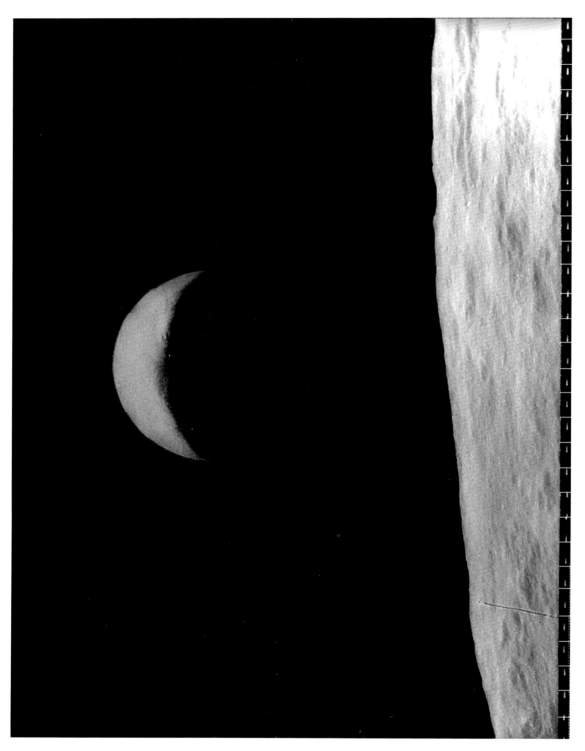

LO1-117H2   Sun Elevation: N.A.   Distance: N.A.

This later photo reveals mottling in the cloud layer. The strong weather patterns of the previous photo have passed to the east with the rotation of the Earth (the sun has set on them). Note the elevation profile of the lunar surface at its horizon. When the line of sight to stars grazes the edge of the Moon, they can be seen to blink.

LO2-196H1   **Sun Elevation: ~20°**   **Distance: ~1520 km**

This provides a close-up view of the southwest sector of Saha (**S**, 99 km) and the southeast sector of the older crater Wyld (see the annotated view of these craters and their context in the photo of LO2-196M). The ejecta of Saha broke down parts of the rim of Wyld (**W**, 93 km); as a result, some of the material (partly from the rim of Wyld and partly ejecta from Saha) landed on the floor of Wyld. The more recent smaller craters in this photo illustrate the progression of shape as size increases, from craters with smooth internal cavities to craters with central peaks.

LO2-196H2   Sun Elevation: 19.86°   Distance: 1519.06 km

The ejecta blanket of Pasteur (**P**, 224 km) covers the rest of this frame showing radial ridges and troughs, especially across the rims of older craters. Pasteur has been classified as pre-Nectarian. Craters that are older than Pasteur and others that are younger can be observed in this frame. Ejecta patterns of some of the younger craters can be seen more clearly than that of Pasteur.

From Pasteur

P

B

LO2-196H3   Sun Elevation: ~20°   Distance: ~1520 km

The floors of Pasteur (**P**, 224 km) and Backlund (**B**, 75 km) are slightly darkened in Clementine albedo data, suggesting the presence of cryptomare, lava covered by a layer of ejecta from nearby craters and basins. The level fill in the proposed Near Side Megabasin is on the left of this photo and the exposed slope of the basin is on the right side. The fine lines outline two fields of secondary craters. The sources of these fields are Saha (99 km) and Pasteur D (36 km), shown in medium-resolution photo LO2-196M.

LO4-009H3   Sun Elevation: 17.81°   Distance: 3011.79 km

This view of northern Australe Basin shows how the lava intruded on the cratered cavity of the Australe Basin, but did not entirely cover the rims of the craters there. This clearly shows that heavy bombardment took place between the time of the basin formation and the time the lava rose. This long time period, from the pre-Nectarian through the Nectarian and Early Imbrian Periods to the Late Imbrian Period, was needed to allow the radioactive elements to heat the basalt to the melting point. Barnard (**B**, 105 km) is the youngest of three craters that overlap. Barndard D (**BD**, 47 km) shows the effect of having been formed on the slope of the Australe Basin: its southern rim, lower than its northern rim, is flooded by Mare Australe.

LO4-009H2   Sun Elevation: 17.81°   Distance: 3011.79 km

The Jenner (J, 71 km) impact happened in the Late Imbrian, during the emplacement of mare. Its ejecta covered the surrounding flows, but its cavity has received additional lava. The northwest sector of the rim of nearby crater Lamb (L, 106 km) is lower than the rest of the rim, due to its having been formed on the slope of the Australe Basin's cavity. Jenner does not show this effect because its target surface was on the level mare flow.

LO4-009H1   Sun Elevation: 17.81°   Distance: 3011.79 km

In this area, the southern ejecta blanket of the Australe Basin, only the bottoms of some of the craters have received mare flows. One of those is Kugler (**K**, Nectarian, 65 km). Anuchin (**A**, 57 km) appears younger than Kugler.

LO1-136H1   Sun Elevation: ~21°   Distance: ~1373 km

Several chains of craters appear in this photo; Catena Gregory (**CG**), slashing across the upper right corner is the largest. Gregory (67 km) is the source of the secondary impactors. It is just to the northwest of this subframe. Although a catena can result from multiple primary impactors, like the Schoemaker-Levy event on Jupiter, they are usually the result of secondary ejecta from impact basins or craters. They occur in the far field, beyond the continuous ejecta blanket that typically extends for one radius beyond a crater's rim. Prager (**P**, 60 km), in the lower right part of this frame, is relatively deep for its size, perhaps because it has impacted thick ejecta just beyond the rim of the Near Side Megabasin.

LO1-136H2   Sun Elevation: 20.65°   Distance: 1373.43 km

Love (**L**, 84 km) and Perepelkin (**P**, 97 km) are very similar in age and size, and they seem to have deposited ejecta on each other's floors. They are likely to have been caused by a pair of primary impactors. Their eastern walls are higher than their western walls, evidence for the Near Side Megabasin, whose estimated rim passes to the east of the center of these craters. This difference between wall heights is typical of craters that span basin rims, such as we will see in Tsiolkovskiy and also in Apollo, which span the rim of the South Pole-Aitken Basin.

LO1-136H3   Sun Elevation: ~20°   Distance: ~1520 km

Shirakatsi (S, 51 km) and Dobrovol'skiy (D, 38 km) are of similar age, as indicated by their degree of degrada- tion, but clearly Shirakatsi is the later impact since it has plowed the rim of Dobrovol'skiy onto its floor.

LO3-121H1   Sun Elevation: ~19°   Distance: ~1533 km

This spectacular view of Tsiolkovskiy's (**T**, 185 km) north-west sector shows the young mare on its floor surrounding the central peak, the rising slope of its basin, and the terraced cliffs of its rim. It has thrown clearly molten ejecta onto the floor of the older crater Fermi (**F**, 183 km) to the northwest. The northern cliffs have exposed a cross section of the target material, probably the rim of the Near Side Megabasin, rising toward the east. In addition to the obvious increase in height of these cliffs, the rise can be seen by following the terraces around to the east. The eastern cliffs of Tsiolkovskiy beyond this frame are even higher (see LO3-121M).

LO3-121H2   Sun Elevation: 19.37°   Distance: 1533.85 km

Ejecta from Tsiolkovksiy (**T**, 185 km) covers the floors of Waterman (**W**, 76 km) and Neujmin (**N**, 101 km). The debris thrown on the floor of Waterman was clearly molten, and appears to have sagged back into Tsiolkovskiy through a destroyed section of the rim. The dark patches on the floor of Waterman are mare flows.

LO3-121H3   Sun Elevation: ~19°   Distance: ~1533 km

There is mare material in the northern part of the floor of Bolyai (**B**, 135 km) but dark areas in the smaller craters to the north and east are probably just shadows. They do not appear to have lower albedo in the Clementine high-sun photos.

LO1-115H1   **Sun Elevation: ~18°**   **Distance: ~1398 km**

One might think that the ridges and troughs here are due to Tamm (**T**, 38 km), but they are mostly due to the much larger crater Chaplygin (**C**, 137 km), out of the subframe to the east. Tamm has a raised fill on its floor that may be molten material deposited by either Chaplygin or the Mendeleev Basin. This area is within the continuous ejecta blanket of each impact feature.

LO1-115H2   Sun Elevation: 18.96°   Distance: 1398.77 km

More of the molten fill in Tamm (**T**, 38 km) is contiguous with similar fill in van den Bos (**vB**, 22 km) and the plains to the south. Striations in the floor and rim of Vil'ev (**V**, 45 km) are radial to Mendeleev. The model of the Near Side Megabasin predicts a scarp running near the eastern edge of Marconi (**M**, 45 km). This scarp is the edge of the layer of ejecta that would be thrown to the west from the Near Side Megabasin. Evidence of this second layer can be seen by following the terraces on the wall of Marconi. The dark patch on the floor of Marconi is not mare, just an artifact of reconstruction of this subframe.

From Dellinger

M

From Tsiolkovskiy

LO1-115H3   Sun Elevation: ~19°   Distance: ~1398 km

The deep gouges in the rim of Marconi (**M**, 45 km) may be due to Dellinger (81 km) to the northwest. The prominent catena that runs diagonally across this frame is radial to Tsiolkovskiy and is probably part of the far field of that crater.

LO2-034H1   Sun Elevation: ~19°   Distance: ~1570 km

This subframe covers the southwestern quadrant of the Freundlich-Sharonov Basin (600 km). The basin's center (and deepest part) is just over the top edge of this photo, north of Virtanen (**V**, 45 km). There is a mare outcrop near the center, delightfully called Lacus Luxuriae (see LO2-034M). Sharonov (**S**, 74 km) lies very near the southwest edge of the basin. Like the other craters in this frame, Sharonov appears to be elliptical because this photo was taken at an oblique angle from the south. Sharonov, essentially free of later craters, is Copernican and Virtanen is Eratosthenian.

LO2-034H2   Sun Elevation: 19.20°   Distance: 1570.57 km

This area is dominated by Valier (**V**, 67 km), Tiselius (**T**, 53 km), and the Copernican Coriolis Y (**CY**, 31 km). Although this general area is far from any larger craters or younger basins, it is influenced by the very old Freundlich-Sharanov Basin to the north and the even older South Pole-Aitken Basin to the south. The proposed Near Side Megabasin would have formed a fairly uniform double layer of ejecta here. There is a full range of ages and sizes of craters in this area, which have fully worked over the terrain. The crater distribution here has approached a steady state.

LO2-034H3   Sun Elevation: ~19°   Distance: ~1570 km

Coriolis (**C**, 78 km) was formed in the later pre-Nectarian Period. It has retained a considerable amount of detail within its basin, but its ejecta field has been badly eroded. The short chains of craters south of Coriolis may be part of the far field of Mandel'sham, to the northwest.

Chapter 7

The Korolev Basin Region

7.1. Overview

The Korolev Basin was an early target of Lunar Orbiter I. The basin is named after Sergei Korolev, the chief designer of the first Soviet intercontinental ballistic missile, the R-7. He also led the design of Sputnik I, the first artificial satellite of Earth. This series of exposures (see Figure 7.5) covers quite a large sample of the far side crust – about 40° in each dimension. An overview of the area is shown in Figure 7.1. As is typical for the highlands on both the near and far sides, the surface is fully battered by craters, one overlapping the other.

This basin is in the highest area of the Moon, a mound that rises about 5 km over the reference geoid (see Figure 7.2). It has been suggested that this mound is ejecta from the nearby South Pole-Aitken basin but although that basin had a contribution, it was a relatively minor one. The major source is, more likely, the proposed Near Side Megabasin, whose antipode is located in the southwest quadrant of the floor of Korolev. In this view, Korolev sits on top of the mound of ejecta from the Near Side Megabasin.

As with all basins, the shape of the original surface where it impacted has modified the Korolev Basin, which has been formed on the curved surface of a mound. This is an illustration of the principle of superposition. As a basin forms in response to an impact, it follows the shape of the target surface. This is obvious in the case of a flat, level target whose only shape parameter is its elevation. For a flat surface with a slope, it is also obvious that the final shape can be estimated by adding the slope to the shape of a typical impact basin. The same principle applies to a surface with curvature, so long as the angle that the slopes make with the horizontal is small. The impact explosion will likely follow the initial target surface and the basin shape will be superimposed on the original target.

Figure 7.3 shows a radial profile of the elevation of the Korolev Basin. The antipode mound has distorted it so that it does not look much like the general model of impact basin radial profiles (see Figure 13.4). The curved line represents an estimate of the shape of the mound prior to the Korolev impact. It has a maximum elevation of 5000 m above the reference geoid of the Clementine database.

After subtracting the quadratic curve from the measured elevation profile, the corrected radial profile is shown in Figure 7.4. Correcting for curvature makes a much better fit to the model of impact basins.

As can be seen, the Korolev Basin is like any other impact basin, once the curvature is removed. The target surface also had an average slope relative to the basin because the basin is a little to the northeast of the antipode. It is not necessary to correct for slope because the process of deriving the radial profile automatically removes it.

There is one internal ring within Korolev, which shows as a bump in Figure 7.4. It can also be seen in the mosaic of Figure 7.2. Also, the center of the floor of Korolev, within the inner ring, has been raised above the level that the model of Figure 13.4 predicts for a fresh basin. This may reflect the deposit of ejecta from the Hertzsprung Basin to the east.

7.2. Photos

Lunar Orbiter Medium-Resolution Frames

The approximate coverage of the medium-resolution photos targeted for the Korolev Basin region is shown in Figure 7.5.

As indicated in Figure 7.5, photos LO1-028M and LO1-030M essentially cover the region (all of the features are annotated in those two photos), and so only those two medium-resolution photos are shown.

Lunar Orbiter High-Resolution Subframes

Photos LO1-030H and LO1-036H overlap and are shown as two mosaics, one for the H1 and H2 subframes and one for the H3 subframes. Photos LO1-038H1, 2, and 3 are shown separately to provide maximum resolution. The locations of LO1-030 and 031 H1, H2, and H3 are marked on the medium resolution photo LO1-030M and the locations of LO1-038H1, H2, and H3 are marked on the medium resolution photo LO1-028M.

Figure 7.1. The Korolev Basin and the northern rim of the South Pole-Aitken Basin (darker surface). This is a mosaic of LO1-028M and LO1-030M (NASA, LPI).

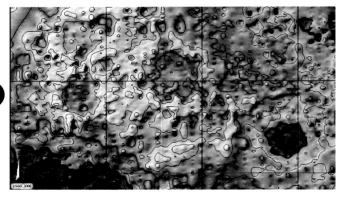

Figure 7.2. This is an elevation map (Topogrd2), from 180° west to 90° west and 30° south to nearly 30° north. Low elevations are dark. Three large basins are, from left to right, Korolev, Hertzsprung, and Orientale. The depression in the lower left corner is the South Pole-Aitken Basin. The contour lines are spaced at 2 km. The highest elevations, shown in white, show the mound of ejecta from the Near Side Megabasin. USGS, http://webgis.usgs.gov/pigwad/maps/the-moon.htm

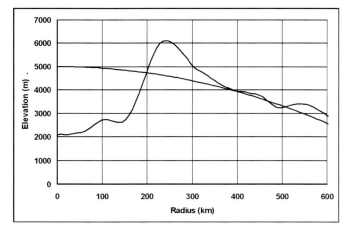

Figure 7.3. The radial profile of elevation for the Korolev Basin (elevations from topogrd2). The *curved line* is an estimate of the elevation of the antipode mound of the Near Side Megabasin.

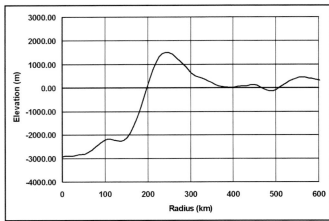

Figure 7.4. The corrected radial profile of depth of the Korolev Basin, derived by subtracting the estimated elevation and curvature of the target surface. Note the clear indication of an inner ring with a radius of 100 km.

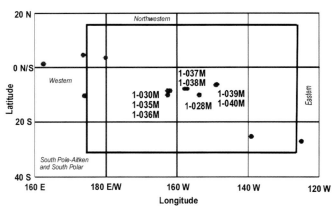

Figure 7.5. This figure shows the principal ground points and approximate area of coverage of the Lunar Orbiter photos targeted at the Korolev Basin region. The terminator imposes the eastern boundary in the later exposures.

Sun Elevation: 23.59°

LO1-030M

Distance: 1300.61 km

Key: Korolev and craters to the west, north of SPA

Nectarian
Congreve, 57 km
Icarus, 98 km
Korolev Basin, 440 km
Zhukovskiy, 81 km

Early Imbrium
Doppler, 110 km
Engelhardt, 43 km

Copernican
Crookes, 49 km

This is the western part of the Korolev Basin Region. The northern rim of the South Pole-Aitken Basin is shown. It would be much more dramatic if the illumination were from the north, which is, of course, impossible. Icarus, De Vries, McKellar, and Bok are examples of craters that have a distinct central peak. The rim of Korolev has been eroded by subsequent bombardment, but the western edge of the rim appears to have formed across a previous crater. Because of the degree of erosion, Korolev has been assigned to an earlier Nectarian period, that is, slightly younger than Nectaris. Lacus Oblivionis and Rumford are dark in Clementine's images and probably have been invaded by lava.

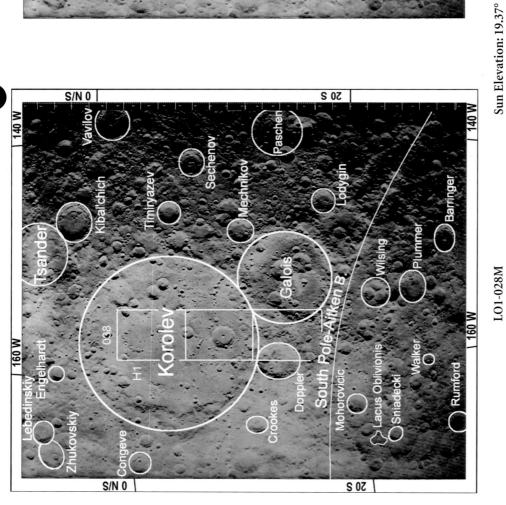

LO1-028M

Sun Elevation: 19.37°

Distance: 1305.28 km

Key: Craters to the east of Korolev, north of SPA

Pre-Nectarian
Galois, 222 km
Paschen, 124 km
Tsander, 181 km

Nectarian
Kibal'chich, 92 km
Mechnikov, 60 km
Sechenov, 62 km
Timiryazev, 53 km

Copernican
Vavilov, 96 km

The smooth parts of the floor of Korolev have been resurfaced in the Nectarian and Early Imbrian periods. Doppler is younger; it has impacted the southern rim of Korolev and thrown part of that rim, plus its own ejecta, onto the floor of the basin. The inner ring of Korolev (see Figure 7.4) is very clear in this photo. Some of the very sharply defined smaller craters may be secondaries from the Orientale Basin to the east–southeast.

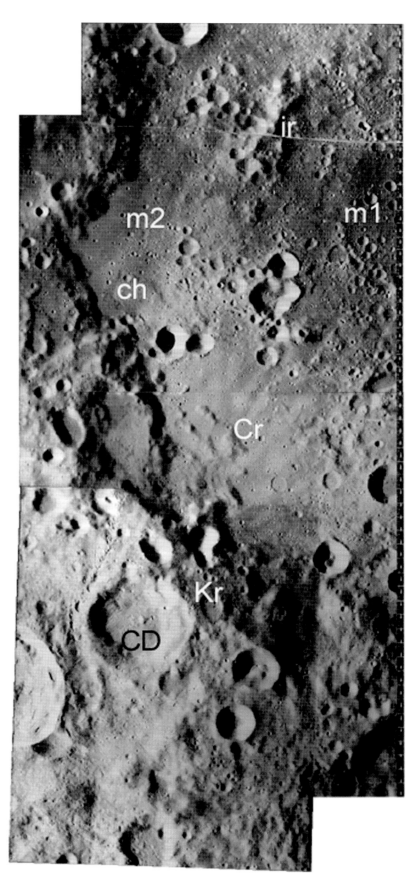

LO1-030H1, H2 (left from top)
LO1-036H1, H2 (right from top)
Sun Elevation: ~22°
Distance: ~1300 km

This photomosaic shows the southwest sector of the Korolev Basin, from its center to its rim (**Kr**). Its central floor is flooded with dark mare (**m1**), bounded by the raised inner ring (**ir**). Some mare material (**m2**) has also intruded into the western sector of the basin between the inner ring and the rim. This young surface is assigned to the Early Imbrian period and the mare in the center to the Nectarian period, later in that long period than Korolev itself, of course. Beyond the inner ring, the basin floor slopes upward to the rim cliffs.

Crookes D (**CD**, 41 km) has impacted the Korolev rim and broken down the rim cliffs. The ejecta blanket and rays (**Cr**) of Crookes have spread light material over Crookes D and onto the floor of Korolev. The eastern half of Crookes (49 km) is just southwest of Crookes D.

Chains of secondary craters (**ch**) radial to the Orientale basin can be seen between the inner ring and the rim.

The southeast corner of the mosaic shows the circular rim of Korolev rising to a sharp maximum elevation and then sloping down into the ejecta blanket, with its hummocks. Some of the young craters on the inner rim appear elliptical because of the steep slope there. This is an illustration of the principle of superposition: the shape of an impact crater is added to the shape of the target surface. If the target surface is sloped, the crater will be slanted accordingly.

LO1-030H3 (left)
LO1-036H3 (right)
Sun Elevation: ~22°
Distance: ~1300 km

This photomosaic continues to the south from the one given before.

The northeast sector of the rim of the South Pole-Aitken Basin (**SPA**, 2500 km) dominates this picture. When this ancient basin was fresh, there were cliffs like those of Korolev. They have been worn away by over 3 billion years of subsequent impacts, including all of the period of the proposed great cataclysm discussed earlier. For example, Doppler (**D**, 110 km) has modified the SPA rim by its impact and ejecta. The high incidence of large craters inside the South Pole-Aitken Basin, relative to their frequency inside of Korolev, is further evidence of its great age.

The brightness of the area between Doppler and the South Pole-Aitken Basin is due to ejecta from Crooks to the northwest.

From Orientale

LO1-038H1   Sun Elevation: 6.3°   Distance: ~1390 km

The sun elevation for this photo is lower than usual. As a result, the detail is exaggerated but more can be seen where the surface is nearly even. This area is in the north-east sector of the floor of Korolev. Mare has flooded and resurfaced this area of the basin floor, within the inner ring; evidence of individual flows can be seen. Most of the craters that were formed on top of the mare surface have had very little erosion. They show a rounded cross section, a sharply peaked rim, and a smooth ejecta blanket. The mare basalt here, like most, was probably exuded after the period of the proposed great cataclysm. There are many secondary craters, some chains of which are radial to Ori-entale. Striations on the northern wall are probably due to Orientale.

From Orientale

LO1-038H2   Sun Elevation: 6.36°   Distance: 1388.34 km

The spectacular crater Korolev M (**KM**, 58 km) has a central peak that is almost perfectly formed. The mare that flooded its floor has hardly been disturbed but craters on its rim indicate that it is from the Nectarian period. The northern rim of this crater has impacted the inner ring of Korolev, probably raising a high sector of that northern rim, which later collapsed onto the floor in response to a small impact that must have come after the mare lava hardened. Other impacts have caused similar events at the southern rim. A catena (chain) of secondary craters from Orientale crosses the picture near the top.

LO1-038H3   **Sun Elevation: ~6.3°**   **Distance: ~1390 km**

In this chaotic area south of Korolev, we see the outer rim of Doppler (**D**, 110 km) superimposed on the outer rim of Galois Q (**GQ**, 132 km), which is superimposed on the outer rim of Korolev (top). Some loss of detail is a consequence of the low sun angle. There is a suggestion of a crater predating both Doppler and Galois Q in the lower half of the picture, a crater whose rim touches each of the other rims, but it may be just a coincidental pattern.

Chapter 8

The South Pole-Aitken Basin and the South Polar Region

8.1. Overview

The South Pole-Aitken Basin is the most prominent feature of the far side. Centered at 54° south latitiude and 169° west longitude, its rim extends from 15° south latitude (just south of the Korolev Basin) to beyond the South Pole of the Moon, at 80°south latitude, over on the near side. The crater Aitken is not especially notable in itself; it simply happens to be near the rim of the basin, opposite the South Pole (also near the rim), and so both Aitken and the pole are convenient references.

Because this enormous basin extends to and beyond the South Pole, the coverage of the entire South Polar Region is included in this chapter.

First Lunar Orbiter and then Apollo photography revealed parts of the mountainous rim of the South Pole-Aitken Basin. Because the basin is so large, it can only be seen in part in individual photos. Several analysts contributed to the description of this basin and the USGS mapped it as a definite feature in 1978. At the Clementine postmission press conference in 1994, Gene Shoemaker enthusiastically presented the Lidar data that confirmed its remarkable size and depth.

Figure 8.1 shows an elevation map centered on the South Pole-Aitken Basin. Note that it is an elliptical basin, indicating that the impactor had a low angle of approach. Its major axis is tilted about 10° toward the west from the north. Multispectral data from Clementine revealed an asymmetrical distribution of heavy elements that suggests that the impactor arrived from the south (Garrrick-Bethell, 2004).

The area of greatest long-range slope on the Moon extends from the deepest part of the Apollo Basin to the mound of Near Side Megabasin ejecta, surmounted by the rim of the Korolev basin. The impact of the South Pole-Aitken Basin into this region of deep, light, porous ejecta may help explain its extraordinary depth, which was even greater before it underwent isostatic compensation.

8.2. Clementine

The Clementine mission, in polar orbit, took brightness data of the entire Moon, including the South Pole-Aitken Basin and the South Polar Regions (see Figures 8.2 and 8.3). This was done in the form of limited-area exposures of a CCD camera from which mosaics were constructed to show very large areas, up to full coverage of the entire Moon.

The South Pole-Aitken Basin is darkened by mare basalt in the floors of the deeper craters and basins within its depression. The South Pole-Aitken Basin was formed early in the pre-Nectarian Period. The incursion by mare basalt was completed in the Late Imbrian Period, which allowed time for the maria here to be extensively overlain with rays and ejecta from craters and basins outside of the region.

Clementine's near-polar orbit was aligned in such a way that the sun was always nearly in the plane of the orbit, so that brightness variations due to topography were essentially lacking near the equator, although there were strong shadows near the poles.

Figure 8.3 shows the South Polar Region, part of which is within the cavity of the South Pole-Aitken Basin. The floors of the deep craters very near the South Pole are always in shadow, which makes them the coldest places on the Moon. The Prospector spacecraft's instruments detected hydrogen in those craters but could not determine whether the hydrogen is frozen or adsorbed or has combined with oxygen to form water ice. Both hydrogen and water arrive on the Moon, the hydrogen in the solar wind and the water from comets, and so many think that both forms may be held in the cold-traps of the polar craters.

8.3. Clementine and Lunar Orbiter Coverage

The approximate coverage of the Clementine images and Lunar Orbiter medium-resolution photos targeted for the South Pole-Aitken Basin and the South Polar Region is shown in Figure 8.4. This combination of coverage provides both a systematic overview of the South Pole-Aitken Basin through Clementine images and good topographic views with Lunar Orbiter photos.

Clementine Mosaics

The image that is presented on page 64 covers the northern part of the South Pole-Aitken Basin, north of 55° south latitude. It is a custom Clementine display, a sinusoidal projection downloaded from the USGS Map-a-Planet web page. This image is the base for a discussion of the general nature of the basin. The South Pole-Aitken Basin is a megabasin; it has within it smaller basins like the Apollo Basin (Figure 8.5).

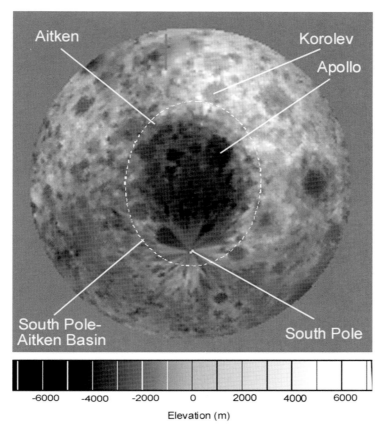

Figure 8.1. Elevation map of the Moon in a Lambert equal area projection (90° range) centered on the South Pole-Aitken Basin. The star pattern around the South Pole is within 12° of the pole, where Clementine was too high in its orbit for the LIDAR instrument to return valid elevation data. NRL, Lambert projection by the present author.

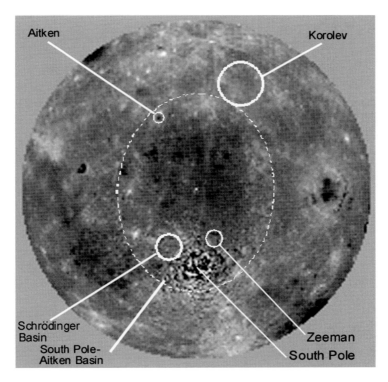

Figure 8.2. This map of Clementine brightness (albedo) is centered on the South Pole-Aitken Basin (Lambert equal area projection, 90° range). Because the Clementine mission took photos at a high sun angle (required for albedo and multispectral imaging), this image shows little topography except at latitudes within 35° of the pole. NRL, Lambert projection by the present author.

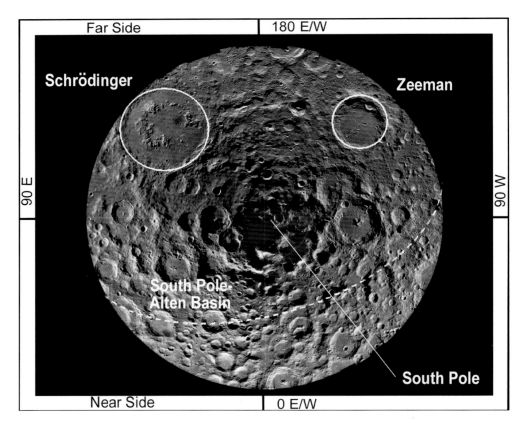

Figure 8.3. This Clementine mosaic shows the South Polar Region from the South Pole to 70° South latitude. The far side is at the top of this photo and it includes the southern portion of the South Pole-Aitken Basin. The NRL assembled this uncontrolled image mosaic, an orthographic projection.

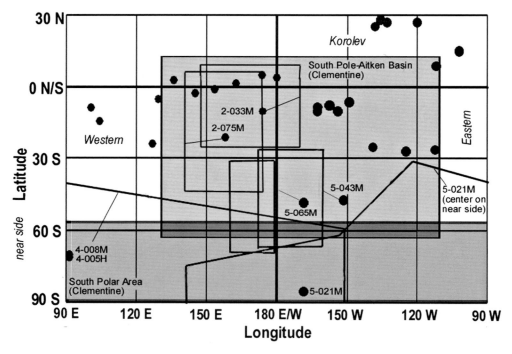

Figure 8.4. This figure labels the approximate area of good coverage of the Clementine and Lunar Orbiter images included in this chapter. Clementine images are shown in gray and Lunar Orbiter medium-resolution frames are outlined, as well as showing their principal ground points.

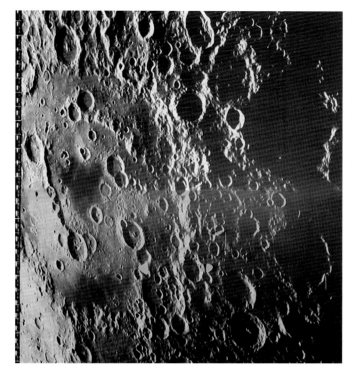

Figure 8.5. The Apollo Basin is on the slope of the South Pole-Aitken Basin, between the flat floor and the northeast rim. Part of LO5-030H1 and H2.

Many of the craters in this area are named for the Apollo astronauts and those who lost their lives in the space shuttle disasters. A special Clementine image is provided to cover this area and each of the craters named in honor of these astronauts is shown there.

The coverage of the South Pole-Aitken Basin continues with a series of custom Clementine images, also in sinusoidal projection, each covering the latitude range from 55° south to 75° south. The far side is covered in four of these images, each 45° in longitude range. The region from 70°

south latitude to the South Pole is covered in an orthographic projection downloaded from the NRL web site. This area shows the permanently shadowed craters that may contain water ice.

Three basins were formed in the South Polar Area during the early periods of heavy bombardment: the Schrödinger Basin, the Sikorsky-Rittenhouse Basin, and the Amundsen-Ganswindt Basin. These basins, each over 300 km in diameter, are so close to each other that they overlap.

Many craters near the South Pole are named after explorers of Earth's South Pole such as Roal Amundsen and Robert Scott. One crater is named as Gene Shoemaker, the geologist who did more than any other to expand our knowledge of impact craters on both the Earth and the Moon.

Lunar Orbiter Medium-Resolution Frames

Lunar Orbiter missions 4 and 5 covered the region of this chapter with a series of oblique photographs. They show a different view of the features than that shown in the Clementine images. The Lunar Orbiter photos cover the latitude range from 30° south to 60° south. As with the Clementine images, the latitude parallels and longitude meridians are marked on one copy of each photo, along with the named craters. An unmarked photo is provided to show the topography clearly. The photos shown are LO2-033M, LO2-075M, LO4-008M, LO5-021M, LO5-065M, and LO5-043M.

Lunar Orbiter High-Resolution Subframes

LO5-005H1, H2, and H3 can be located on LO4-008M. Subframes LO5-021H1, H2, and H3 are assembled into a single frame that is shown on LO5-021M. A full set of subframes are shown for LO2-075 and LO2-033, and can be located by reference to their respective medium-resolution photos.

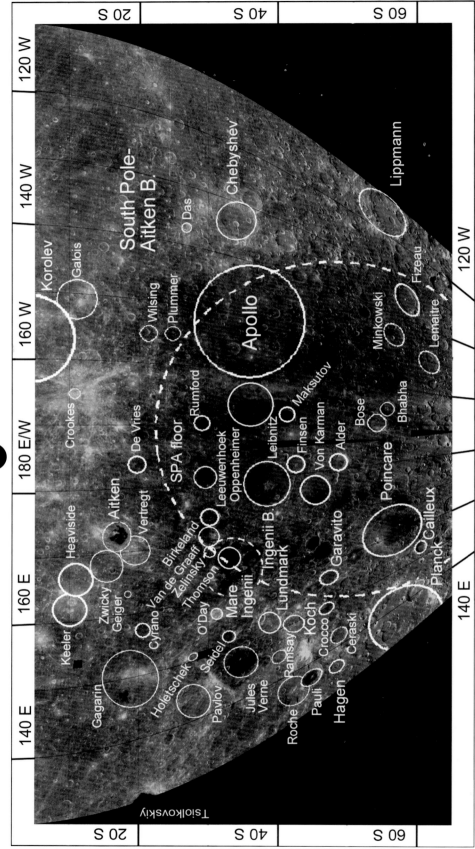

Clementine: Northern South Pole-Aitken Region

This Clementine mosaic, in a sinusoidal projection, is in the northern portion of the South Pole-Aitken Basin (SP-A Basin). The dashed line is near the rim of the basin. The Apollo Basin was named to honor the Apollo missions that explored the Moon, including their photography of the far side, taken from the Apollo Command Modules, starting with Apollo 10. Many craters in and near the Apollo Basin are named after Apollo astronauts and those who have lost their lives in the two Space Shuttle disasters (see page 66).

Key: Features inside or near the SP-A Basin (Early Imbrian or older)

Pre-Nectarian
Apollo Basin, 505 km
Ingenii Basin, 560 km
Jules Verne, 143 km
Lippmann, 160 km
Planck, 314 km
Poincare, 319 km
SP-A Basin, 2500 km
Vertregt, 187 km

Nectarian
Bose, 91 km
Chebyshev, 178 km
Lemaitre, 133 km
Korolev Basin, 440 km
Leeuwenhoek, 125 km

Von Karman, 180 km
Zwicky, 150 km

Oppenheimer, 208 km
Roche, 160 km
Van De Graaff, 233 km

Early Imbrian
Alder, 77 km
Pauli, 84 km
Rumford, 61 km

Clementine: Northern South Pole-Aitken Region

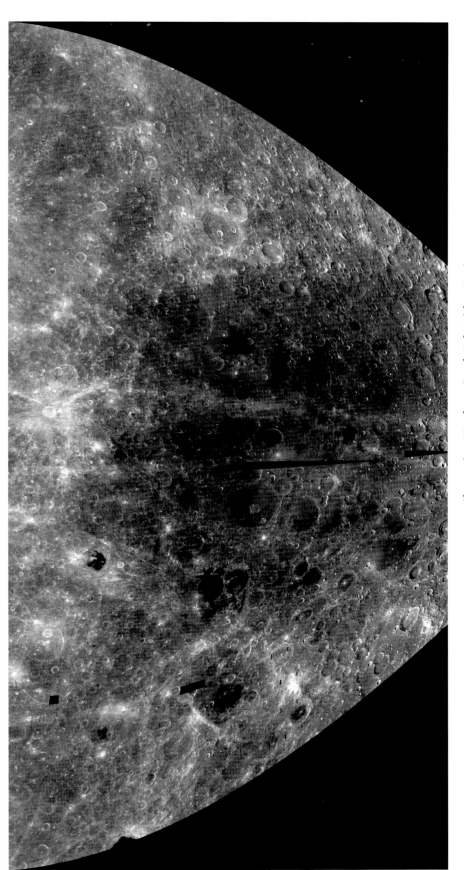

The interior of the South Pole-Aitken Basin is darker than the surrounding area because the large craters and basins within it have been flooded with mare material that has been widely distributed by subsequent impacts. The large, fainter overlying rays come from the Copernican crater Jackson (71 km, 22.7°N, 163.1°W). This basin is the deepest on the Moon – 6.8 km deep.

Key: Features inside or near the SP-A Basin (Late Imbrian or younger)

Late Imbrian
Aitken, 135 km
De Vries, 59 km
Geiger, 54 km
Holetscheck, 38 km
Leibnitz, 245 km
Zelinsky, 53 km

Eratotosthenian
Birkeland, 82 km
Finsen, 72 km

Copernican
Crookes, 49 km
Das, 38 km
O'Day, 71 km

The South Pole-Aitken Basin 65 **and the South Polar Region**

Clementine: Apollo Basin Area

Key: Early Imbrian or older

Pre-Nectarian
Apollo Basin, 505 km

Nectarian
Barringer, 68 km
Bose, km
Oppenheimer, 208 km

Early Imbrian
Borman, 50 km
Rumford, 61 km

Apollo astronauts who are honored by naming craters in and near the Apollo Basin are Edwin E. Aldrin, Jr., William A. Anders, Neil A. Armstrong, Frank Borman, Roger B. Chaffee, Michael Collins, Virgil I. Grissom, James A. Lovell, Jr., and Edward H. White, II. Chaffee, Grissom, and White died in the pad fire of Apollo 1.

Those honored who died in the launch disaster of Challenger are Gregory B. Jarvis, Sharon Christa McAuliffe (the civilian school teacher), Ronald Erwin McNair, Ellison Shoji Onizuka, Judith Resnick, Francis Richard Scobee, and Michael John Smith.

Those honored who died in the Columbia re-entry disaster of Space Shuttle Columbia are Michail Phillip Anderson, David McDowell Brown, Kalpana Chawla, Laurel Blair Salton Clark, Rick Douglas Husband, William Cameron McCool, and Ilan Ramon.

Key: Late Imbrian or younger

Late Imbrian
Dryden, 51 km
Maksutov, 83 km
White, 39 km

Eratosthenian
Bok, 104 km
Lovell, 34 km

If the Apollo Basin (about 500 km in diameter) had formed on the near side of the Moon rather than the far side, it would have been entirely flooded with mare. However, the thicker crust on the far side has limited the amount of flooding.

The dark, roughly circular spot in the center of this image is the mare flooding the inner ring of the Apollo Basin. The arc of mare in the southwestern portion of the trough between the inner ring and the rim has probably risen just in that sector because the floor of the South Pole-Aitken Basin, which was the target surface for the Apollo impactor, is lower there. The edge of the ejecta blanket of the Apollo Basin can be seen as a shadow near the left edge of this photo.

The South Pole-Aitken Basin 67 and the South Polar Region

Southwest Far Side: Western Sector

Key

Pre-Nectarian
Planck, 314km

Nectarian
Sikorsky-Rittenhouse Basin, 310km

Early Imbrian
Schrödinger Basin, 320km

Late Imbrian
Hale, 83km

Erotosthenian
Grotrian, 37km

Mare Basalt has pressed upward here in the Eratosthenian Period, and beyond the mare are dark plains units, another means of resurfacing. These may be caused by fire fountains or may be due to mare that has flooded and receded, leaving a very thin layer of basalt whose darkness has been dimmed by the gardening effect of later bombardment.

These units appear to have modified the ejecta blankets of Planck and Schrödinger. Vallis Schrödinger and Vallis Planck extend beyond the ejecta blanket of Schrödinger as strings of small secondary craters, possibly accompanied by finely divided material. Both valleys are more prominent in the low east–west lighting of the Lunar Orbiter photos of this area.

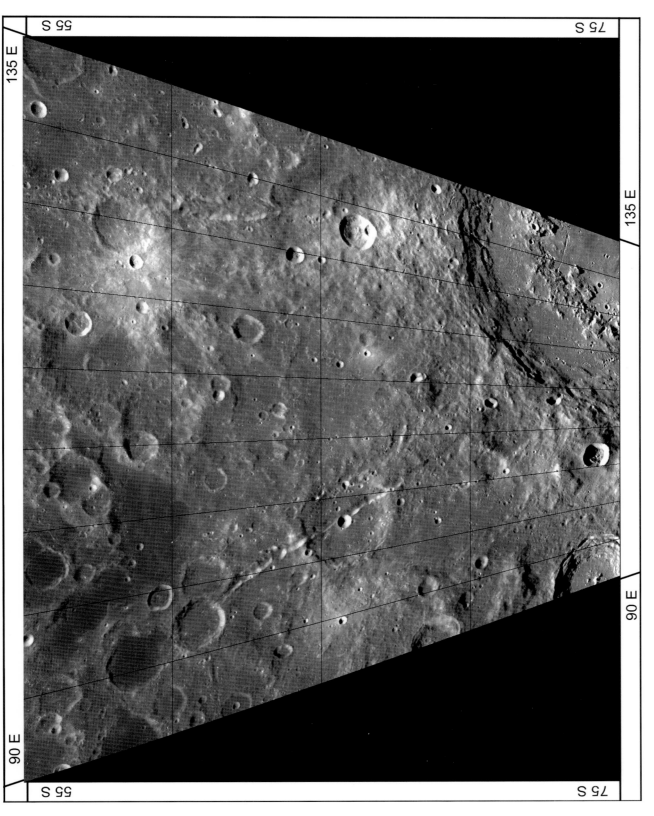

Southwest Far Side: Western Sector, 55° S to 75° S, 90° E to 135° E

The South Pole-Aitken Basin 69 and the South Polar Region

Southwest Far Side: Central Sector

Key

Pre-Nectarian
Planck, 314 km
Poincare, 319 km

Early Imbrian
Schrödinger Basin, 320 km

Late Imbrian
Lyman, 84 km
South Pole-Aitken Basin, 2500 km

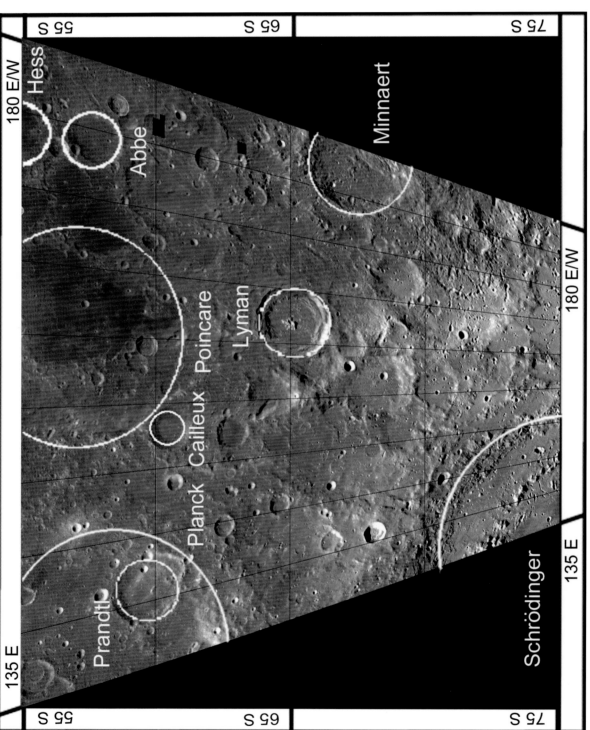

55 S
65 S
75 S

135 E
180 E/W

Hess
Abbe
Minnaert
Poincare
Lyman
Prandtl
Planck
Cailleux
Schrödinger

This area is entirely within the South Pole-Aitken Basin. The mare basalt within Poincare, like nearly all such material in the South Pole-Aitken Basin, has flooded the floor of Poincare in the Late Imbrian Period, erasing signs of any secondary craters from the Orientale basin that may have once formed.

The outlines of craters buried by the ejecta blankets of Schrödinger and Poincare can still be seen. They may have formed either in the pre-Nectarian Period or early in the Nectarian Period.

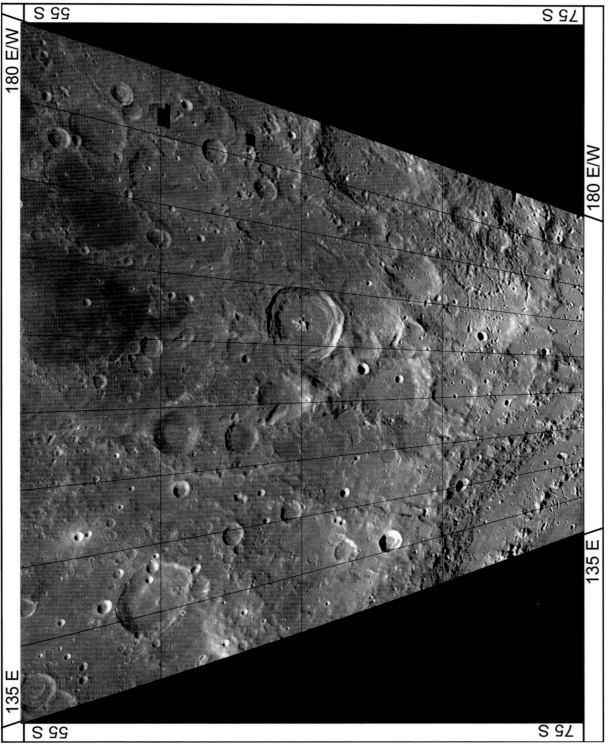

Southwest Far Side: Central Sector, 55° S to 75° S, 135° E to 180° E/W

The South Pole-Aitken Basin 71 and the South Polar Region

**Southeast Far Side:
Central Sector**

Key

Pre-Nectarian
Minnaert, 125 km

Nectarian
Bose, 91 km
Lemaître, 91 km
Numerov, 113 km
Zeeman, 190 km

Early Imbrian
Dawson, 45 km

Late Imbrian
Antoniadi, 143 km
Fizeau, 111 km

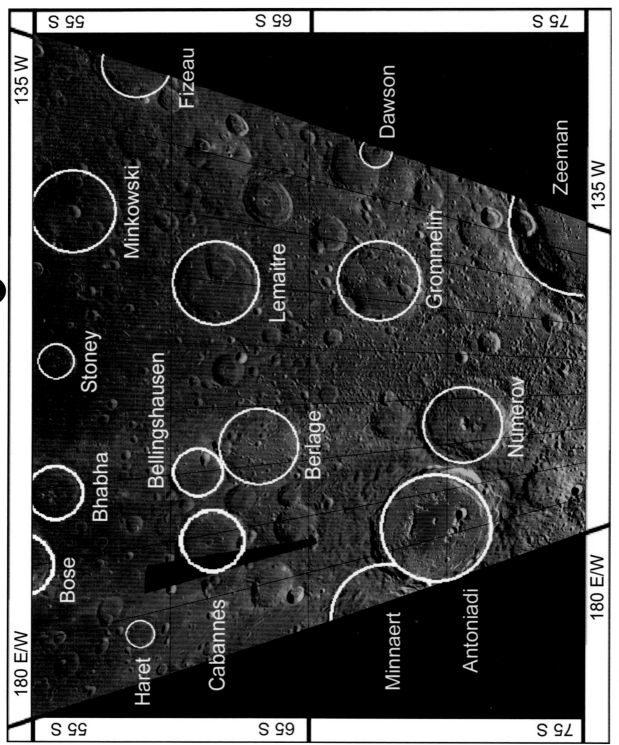

This area of the South Pole-Aitken Basin has less fresh mare than the terrain to the west. The transition between the flat floor of the South Pole-Aitken Basin and the rise to the rim occurs at about 70° south latitude, passing through Zeeman. This transition of material may be the cause of the curious pattern of Zeeman's northwest rim.

Antoniadi has impacted on the rims of Minnaert and Numerov. The condition of their walls and ejecta blankets signifies that Numerov is quite a bit older than Antoniadi, and Minnaert much older than Numerov

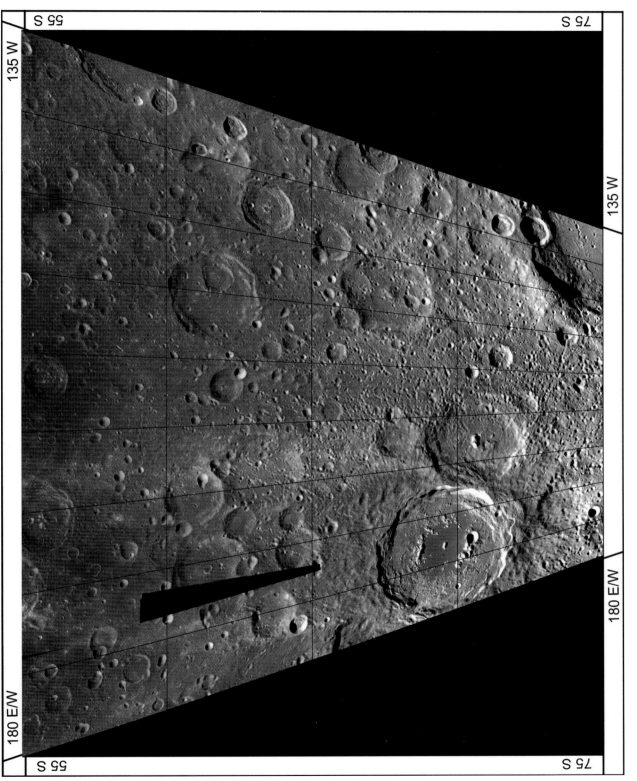

55 S

75 S

135 W

135 W

180 E/W

180 E/W

55 S

75 S

Southeast Far Side: Central Sector, 55° S to 75° S, 135° W to 180° E/W

The South Pole-Aitken Basin **and the South Polar Region**

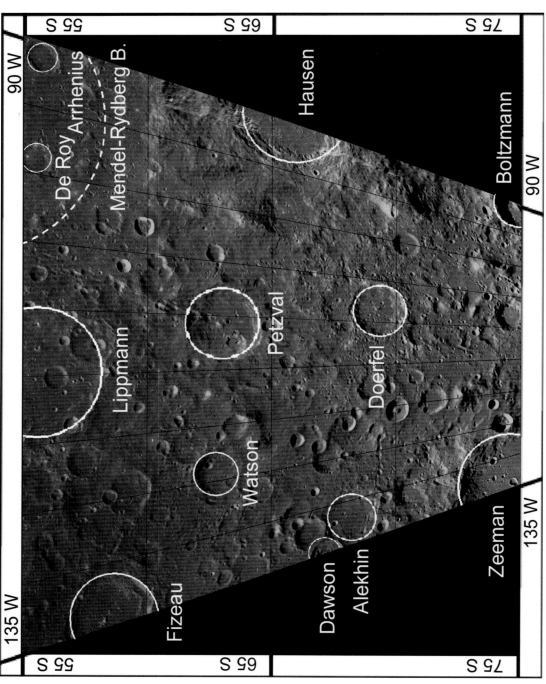

Southeast Far Side: Eastern Sector

Key

Pre-Nectarian
Lippmann, 160 km

Nectarian
Mendel-Rydberg Basin, 630 km
Petzval, 90 km

Early Imbrian
Dawson, 45 km

Late Imbrian
De Roy, 43 km
Doerfel, 68 km
Fizeau, 111 km

Eratosthenian
Hausen, 167 km

The crater Hausen has impacted the eastern rim of the South Pole-Aitken Basin, throwing an unusually deep ejecta blanket onto the floor of the basin. Long before Hausen struck, the Mendel-Rydberg Basin had left its ejecta, now buried by a tongue of Orientale ejecta, which ended on the floor of Petzval.

To review the sequence, first the South Pole-Aitken Basin formed its floor, sloping up to its rim on the right side of the image. Then the impactor of the Mendel-Rydberg Basin struck, throwing an ejecta blanket about twice the basin's radius. Petzval struck that, but so long ago, there is no sign of its own ejecta blanket. The Orientale Basin threw its tongue of ejecta across Mendel-Rydberg and the rim of the South Pole-Aitken basin. Finally, Hausen struck that rim and knocked some of it down the sloping floor of the South Pole-Aitken Basin. The darker area around Fizeau is where the floor levels out. The floor of Fizeau is about 2 km below the rim of Hausen.

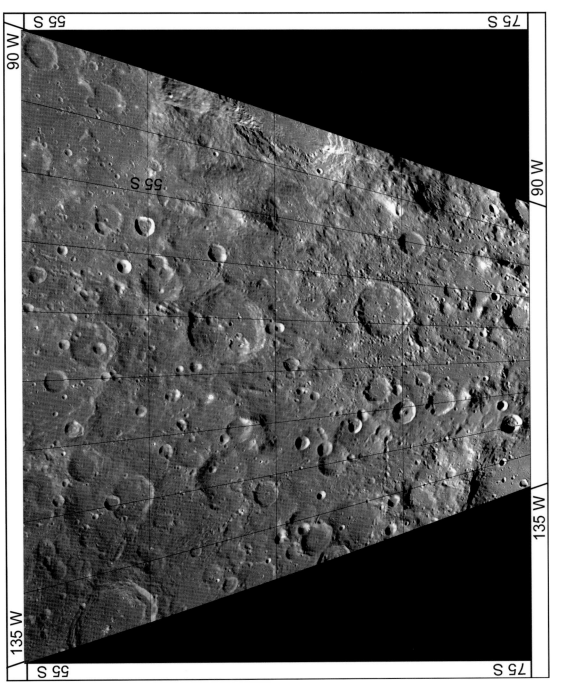

Southeast Far Side: Central Sector, 55° S to 75° S, 90° W to 135° W

The South Pole-Aitken Basin 75 and the South Polar Region

South Polar Cap 70°
South to the South Pole

Key

Pre-Nectarian
Amundsen-Ganswindt Basin, 355 km
Boguslawsky, 97 km

Nectarian
Amundsen, 101 km
Demonax, 128 km
Newton, 78 km
Sikorsky-Rittenhause Basin, 310 km
Zeeman, 190 km

Early Imbrian
Schrödinger Basin, 320 km

Late Imbrian
De Forest, 57 km
Hale, 83 km
Le Gentil, 128 km
Schoenberger, 85 km

Eratosthenian
Moretus, 111 km

Near Side

180 E/W

90 W

90 E

0 E/W

Far Side

Zeeman
De Forest
Brashear
Schröödinger B.
Rittenhouse
Ganswindt
Hale
Sikorsky-
Rittenhouse B.
Neumayer
SPA Rim
Boussingault
Idelson
Amundsen-
Ganswindt B.
Sverdrup
Faustini
Shackleton
Amundsen
Scott
Demonax
Boguslawsky
Schomberger
Simpellus
Short
Newton
Shoemaker
Delgeriache
Drygalski
Casatus
70
Le Gentil
SPA Rim
Ashbrook
80

This South Polar Cap base image is an orthographic projection from the pole to 70° south latitude. It is derived from Clementine visual data from the NRL. This image, processed by USGS, is from the JPL web page. The far side is in the upper half of this image; the near side is on the lower half. The craters listed by period in the key above are all on the far side. The rim of the South Pole-Aitken Basin extends over the pole to the far side, as shown by the dashed line.

South Polar Cap

The southern rim of the South Pole-Aitken Basin (dashed line) extends beyond the South Pole into the near side of the Moon. The Schrödinger Basin has a nearly complete single inner ring. As is typical of such rings, it has a radius of approximately half the radius of the outer rim. Its smaller companion Zeeman shows no evidence of an inner ring or central peak. Drygalski, even smaller, has a central peak as do Short, Demonax, and others in this size range. It is interesting that, at least in this particular area, there is a transition from central peak to a totally flat floor to a single inner ring as the diameters grow from 149 to 190 to 320 km. The impact of Drygalski on Ashbrook shows the power of the expanding crater. Drygalski has broken through the rim of Ashbrook and the velocity imparted to the material of both rims carries it onto the floor of Ashbrook.

The lighting angle is constant with azimuth around the pole. The depth of craters outside the rim of the South Pole-Aitken Basin is greater than that of those within it, as indicated by the greater degree of shadowing. Craters in ejecta are relatively deeper than those that are formed within basin floors.

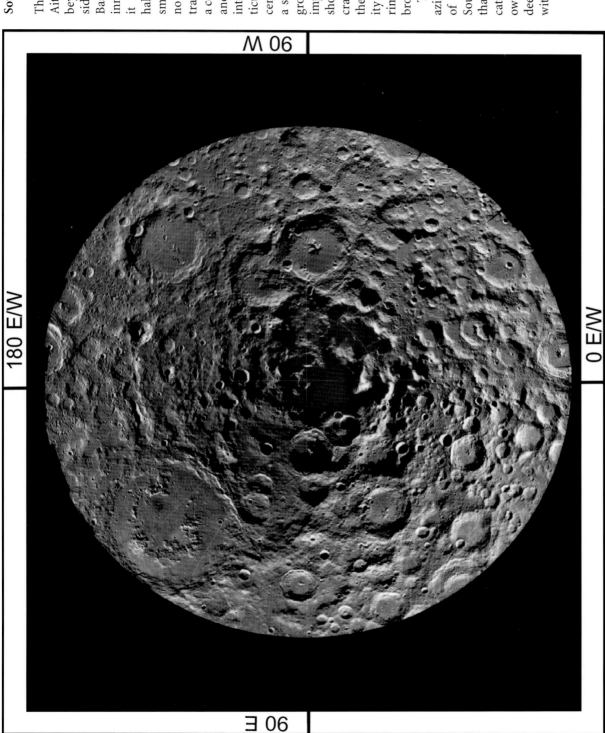

180 E/W

90 W

90 E

0 E/W

Distance: 2753.15 km

Pre-Nectarian
Jeans, 79 km
Minnaert, 125 km
Planck, 314 km
Poincare, 319 km

Nectarian
Amundsen, 101 km
Numerov, 113 km
Zeeman, 190 km

Early Imbrian
Schrödinger Basin, 320 km

Late Imbrian
Antoniadi, 143 km
Hale, 83 km
Schoenberger, 85 km

Eratosthenian
Moretus, 111 km

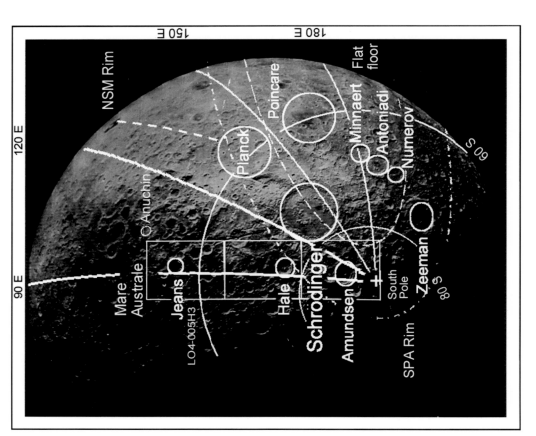

LO4-008M

The Lunar Orbiter 4 spacecraft was nearly over the South Pole when it took this photo, looking to the South Pole-Aitken basin. Note the strong troughs radiating from Schrödinger (the tip of one of these grooves is shown in LO4-005H3). Dashed lines show the estimated rims of the South Pole-Aitken Basin and the Near Side Megabasin and the edge of the flat floor of the South Pole-Aitken Basin. In both cases, the rim slopes are so low and the rims are so eroded that they are not apparent in this photo.

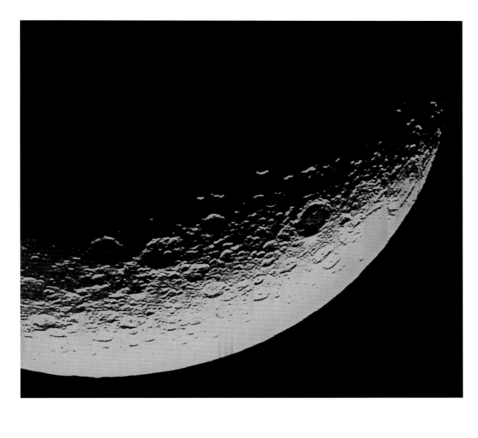

LO5-021M

Distance: 3917.49 km

LO5-021H

It helps to stand on your head to understand this photo. The spacecraft was over the southwestern near side; the camera captured Zeeman and Schrödinger by shooting over the South Pole. The South Pole-Aitken Basin extends beyond Zeeman and Schrödinger through Numerov and Antoniadi. A very nice view of Schrödinger and the South Pole is in LO5-021H.

The South Pole-Aitken Basin 79 and the South Polar Region

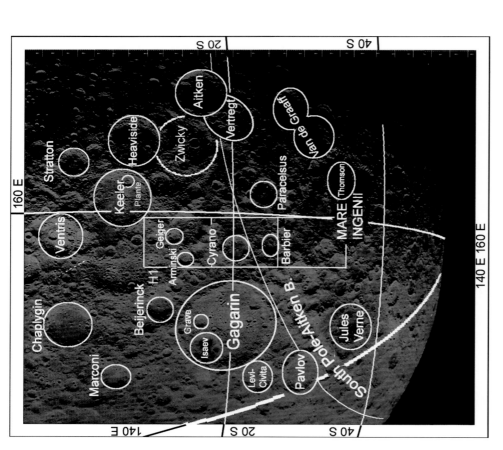

LO2-075M

Distance: 1542.58 km

Key

Pre-Nectarian
Cyrano, 80 km
Gagarin, 265 km
Heaviside, 165 km
Ingenii Basin, 560 km
Jules Verne, 143 km
Vertragt, 187 km
Zwicki, 150 km

Nectarian
Chaplygin, 137 km
Pavlov, 148 km
Stratton, 70 km

Early Imbrian
Keeler, 160 km
Marconi, 73 km

Late Imbrian
Aitken, 86 km
Geiger, 34 km
Holetschek, 38 km

Eratosthenian
Plante, 37 km

The large crater Gagarin has had its floor resurfaced in very ancient times, according to crater counts on its surface. Also the sector of its rim that faces the South Pole-Aitken Basin is lower than the far rim. It is just speculation, but could it have been formed before the South Pole-Aitken Basin and been partly buried by the ejecta blanket of the South Pole-Aitken Basin?

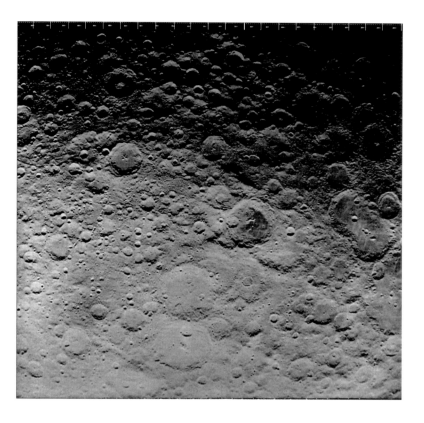

Distance: 1454.85 km

LO2-033M

The Keeler-Heaviside Basin is clearly very old. Although the terrain of its floor is a little softened, it is still nearly covered with craters of rather large size (including Heaviside) that are themselves heavily cratered. Keeler is a relatively recent arrival. Icarus is quite near the Korolev Basin (Nectarian, 440 km), on the edge of that basin's ejecta blanket.

Key, features outside of the South Pole-Aitken Basin

Pre-Nectarian
Cyrano, 80 km
Heaviside, 165 km
Keeler-Heaviside Basin, 780 km
Racah, 63 km

Nectarian
Coriolis, 78 km
Dewar, 50 km
Stratton, 70 km
Vening Meinesz, 87 km

Early Imbrian
Daedalus, 93 km
Keeler, 160 km

Late Imbrian
Aitken, 86 km

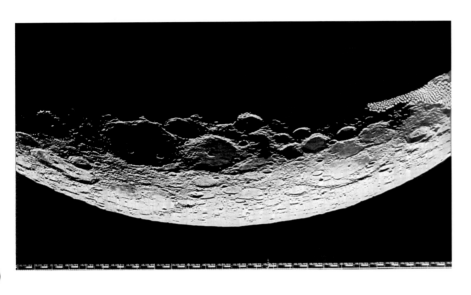

Distance: 1331.39 km

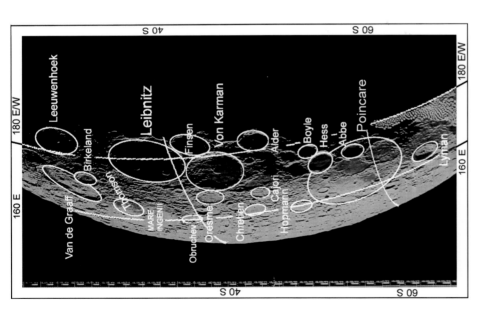

LO5-065M

This photo shows the western South Pole–Aitken Basin. Note the shallowness of the large craters and basins on the floor of the South Pole–Aitken Basin such as Poincare. Van de Graaff is not an elongated crater; it is an overlapping pair of craters. Since neither of its craters appears to have impacted the wall of the other, the two impacts may have been virtually simultaneous, a pair of fellow travelers arriving within at most minutes of one another.

Key

Pre-Nectarian
Abbe, 66 km
Boyle, 57 km
Hess, 88 km
Leibnitz, 245 km
Lyman, 84 km
Oresme, 76 km

Nectarian
Hopmannn, 88 km
Leeuwenhoek, 125 km
Van De Graaff, 104 km

Poincare Basin, 340 km
Von Karman, 180 km

Early Imbrian
Alder, 77 km

Eratosthenian
Birkeland, 82 km
Finsen, 72 km

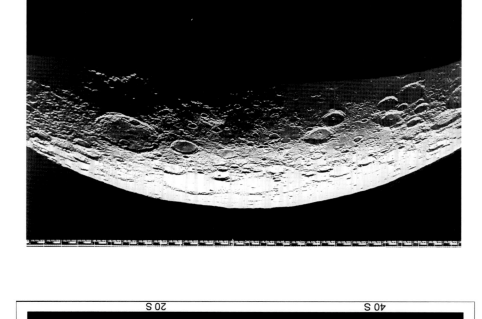

LO5-043M

Distance: 1334.59 km

Key

Pre-Nectarian
Abbe, 66 km
Cabannes, 80 km
Hess, 88 km
Leibnitz, 245 km
Minnaert, 125 km
Nishina, 65 km
Von Karman, 180 km

Nectarian
Bellingshausen, 63 km
Bhabha, 64 km
Bose, 91 km
Oppenheimer, 208 km

Early Imbrian
Alder, 77 km

Late Imbrian
Maksutov, 83 km

Eratosthenian
Finsen, 72 km

This photo shows the central portion of the South Pole-Aitken Basin. The terminator is near the 160°West meridian of longitude. The center of the South Pole-Aitken Basin (54.2°South and 168.7°West) is within crater Bose. The Apollo Basin lies just to the northeast of Oppenheimer. Oppenheimer is judged to be Nectarian in age because it retains the terraces on its rim, yet is heavily cratered.

The South Pole-Aitken Basin 83 **and the South Polar Region**

LO4-005H3   Sun Elevation: 8.6°   Distance: 3512.53 km

This high-resolution frame shows the floor of the Australe Basin, with its craters partly flooded with mare, thinning toward the main ring of the basin (**Amr**). Only the central depressions of the craters near the rim are flooded, where the sloping floor of the Australe Basin rises. The tip of Vallis Schrödinger (**VS**) is resolved here into a chain of secondary craters. Yet note that the far rim of the crater at its tip has also been grooved and debris from that rim has been thrown on the floor of the crater. Although this destruction has been attributed to a horizontal projectile, it may have been caused by secondaries traveling in a higher trajectory, about 45°, and so they would have both vertical and horizontal components of momentum. The horizontal component could produce effects similar to those of a projectile at a low angle. Scale can be determined from the diameter of pre-Nectarian Jeans (**J**, 79 km).

From Schrödinger/

SPA rim /   From Demonax /

LO4-005H2   Sun Elevation: 8.6°   Distance: 3512.53 km

Hale (**H**, 81 km) is young, judging by the degree of detail that is preserved in its terraced wall and the complex structure of its inner ring. Its floor, rim, and ejecta blanket are nearly free of subsequent cratering. However, its rays have completely faded, and so it appears to be Eratosthenian. An arc of the rim of the South Pole-Aitken Basin (2500 km) runs near Hale (see LO4-008M). The large-scale ridges and troughs in the lower right come from the Schrödinger Basin. The finer striations at right angles to the pattern from the Schrödinger Basin are probably from the near side crater Demonax (128 km).

LO4-005H1   Sun Elevation: 8.6°   Distance: 3512.53 km

The South Pole is at the rim of Shackleton (**Sk**, 19 km). Even within the terminator, the floors of some of these craters are in shadow. Because the Moon's axis is only 1.5° from a right angle to the Earth's path around the Sun, some of these crater floors are never struck by sunlight and must be extremely cold. Instruments on the Prospector spacecraft have detected hydrogen concentrations here, in a form that must be stable. Is it ice or simply hydrogen? The rim of the Amundsen-Ganswindt Basin (**AG**, 355 km) is very near the pole, running from the south rim of Amundsen (**A**, 101 km) beyond Ganswindt (**G**, 74 km).

Shoemaker (**Sm**, 50.9 km) is in the shadow that is left of the South Pole. It was named after Gene Shoemaker, pivotal investigator of the geology of impact craters and the founder of the USGS Astrogeology Branch. A portion of Gene's ashes, aboard the Lunar Prospector spacecraft, was sent to impact this crater at the end of its mission.

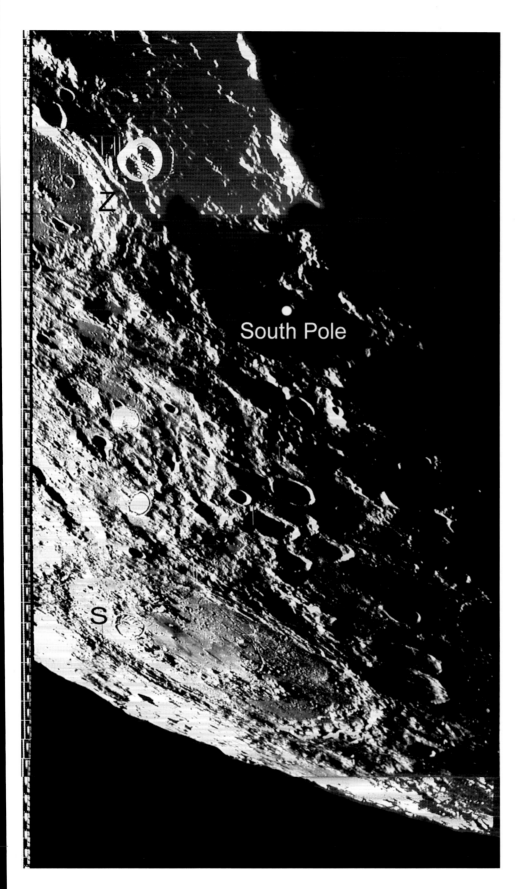

As shown in LO5-021M, this photo was shot over the South Pole (or under, if you like). Consequently, this oblique view of Schrödinger (**S**, 312 km), over the pole, has its south rim above its north rim and its east rim left of its west rim. This view is a mosaic of parts of the LO5-021H subframes. This oblique view of Schrödinger clearly shows its internal ring. Note also that its floor, inner ring, and rim are relatively free of subsequent craters, an indication of a young age. Crater counts indicate that it is of the Early Imbrian Period.

Zeeman (**Z**, 142 km)) is older than Schrödinger; the ejecta blanket of Schrödinger dominates the ejecta field of Zeeman.

LO2-075H1   Elevation: 19.26°   Distance: 1542.58 km

Geiger (**G**, 34 km) has a fresh, sharp look in comparison with its Nectarian neighbor Aminski (**A**, 26 km) and an ejecta blanket that does not have any sizable craters except a few secondaries. Since Geiger has been classified as Late Imbrian, these secondaries cannot have come from any basin. They could have come from Aitken, also Late Imbrian. The "swoosh" is a development artifact.

LO2-075H2   **Elevation: 19.26°**   **Distance: 1542.58 km**

Cyrano (**C**, 80 km) is quite deep. The resurfaced floor must have only a thin layer of mare basalt. The heavy cratering of its rim and the condition of its impact blanket indicates a pre-Nectarian impact.

LO2-075H3   Elevation: 19.26°   Distance: 1542.58 km

This is our best look at Mare Ingenii (**MI**), filling the main ring of the Ingenii Basin (**IB**, 560 km). The young crater peaked on the rim of the Ingenii Basin is O'Day (**OD**, 71 km). Although O'Day has been classified as Copernican, its rays have faded, except on the dark mare, and it may someday be reclassified as Eratosthenian.

DU

LO2-033H1   Elevation: 20.20°   Distance: 1454.84 km

Dark mare basalt has seeped into the very old unnamed crater near the top of this image, above Daedelus U (**DU**, 20 km).

The mare must be very thin here, just barely resurfacing
parts of the crater floor.

LO2-033H2   Elevation: 20.20°   Distance: 1454.84 km

The striations (**st**) here are coming from Aitken (see the next page). These radial patterns of ejecta have marked the rims and floor of Aitken Y (**AY**, 35 km) and Daedalus R (**DR**, 41 km).

LO2-033H3   **Elevation: 20.20°**   **Distance: 1454.84 km**

Aitken (**A**, 135 km) is very deep, a result of its impactor striking the fractured, porous ejecta in the rim of the South Pole-
Aitken Basin. Aitken shows excellent detail in the terraces of
its wall, ejecta blanket, and far field (previous page). Yet it has
received significant cratering on its rim and ejecta blanket. It
is assigned to the Late Imbrian Period.

Chapter 9

The Northwestern Far Side Region: The Moscoviense Basin

9.1. Overview

This region covers that part of the northwestern far side between the East Limb, past the 180° meridian, all the way to 150° west longitude. It covers from 20° north to 60° north latitude.

This area is entirely covered with craters of all sizes and a few basins. The upper layers of the surface have been deposited by ejecta from several basins. Most recently, the Humboldtianum Basin and the Moscoviense Basin (Figure 9.1) arrived in the Nectarian Period, the time range when most of the large craters struck as well. In the pre-Nectarian era, Freundlich-Sharanov and Birkoff left their deposits. Below those layers is a layer from the South Pole-Aitken Basin that is at least 100 m thick.

The proposed Near Side Megabasin would have deposited the foundation layer of ejecta over most of this region. The rim of the Near Side Megabasin runs along the 130° east meridian. Its ejecta blanket covers all the Northwestern Far Side Region east of the rim.

The deep layers of ejecta material that have been reworked by going through the impact process repeatedly have resulted in a surface that is very uniform in element composition. Except for Mare Moscoviense, there are only a few small patches of mare material.

9.2. The Moscoviense Basin

The Moscoviense Basin is the most prominent of the basins that have extensively modified this area. The heavy ejecta blanket of the Moscoviense Basin has covered the central part of this region but has been covered in turn by ejecta from the Humboldtianum Basin (to the north) and the Mendeleev Basin (to the south).

The images of the Moscoviense Basin in Figures 9.1 and 9.2 are taken from oblique medium-resolution photos taken on different orbits, with different sun elevation angles. The main ring, 445 km in diameter, can be seen in both images. The inner ring is 210 km in diameter.

Chains of craters formed by secondary impactors ejected from Moscoviense extend into the outer reaches of its ejecta blanket, surviving ages of erosion from later impacts. Long after the basin was formed, the floor of Moscoviense was flooded with mare, covering the impact features that were formed in the interim and creating a new surface for the lower rate of smaller impacts that followed.

9.3. Photos

Lunar Orbiter Medium-Resolution Frames

The approximate coverage of the medium-resolution photos targeted for the Western Far Side Region is shown in Figure 9.3. Lunar Orbiter 5 took these pictures in a systematic series, with successive medium-resolution frames overlapping. The pictures were taken at high altitude and obliquely because the mission profile was compromised between completion of the medium-resolution coverage of the far side and the high-resolution coverage of the candidate landing sites for the later Apollo landings that were to explore sites of special scientific interest. Photos LO5-085M and -103M have been combined in a mosaic, as have L05-158M and -163M.

The medium-resolution photos are shown starting on page 96. All of the medium-resolution photos are shown before the high-resolution frames, to avoid breaking the continuity of contiguous coverage of the medium-resolution photos.

Lunar Orbiter High-Resolution Subframes

There are three high-resolution subframes for each of the medium-resolution frames that are labeled in Figure 9.1. Their locations are shown precisely on the corresponding medium-resolution frames. Features that are mentioned in the notes with each subframe are identified with code letters.

Figure 9.1. The Moscoviense Basin, seen at a low sun elevation angle. The entire main ring (445 km) is prominent. There is most of an inner ring but mare covers its northeast segment. Part of LO5-124M.

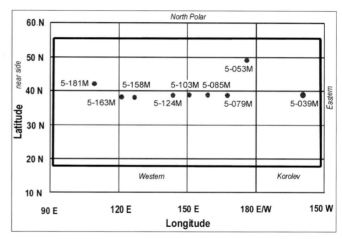

Figure 9.3. This figure labels the principal ground points and approximate total area of coverage of the medium-resolution Lunar Orbiter photos included in this chapter.

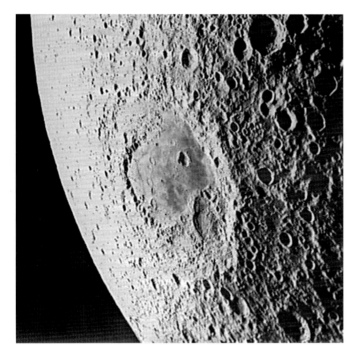

Figure 9.2. The Moscoviense Basin, seen from an earlier orbit at a higher sun angle. Part of LO5-103M.

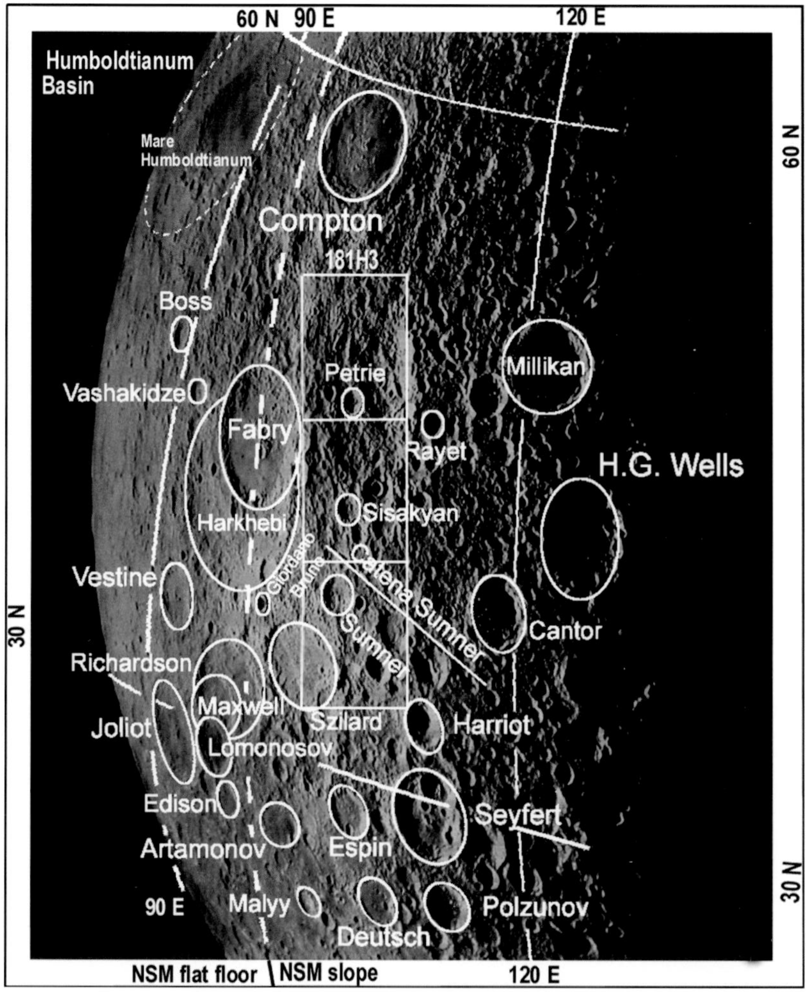

Distance: 1374.79 km

Sun Elevation: 13.32°

LO5-181M

Key

Pre-Nectarian
Fabry, 184 km

Harkhebi, 237 km
Joliot, 164 km
Szilard, 122 km

Nectarian
Humboldtianum Basin,
600 km

Early Imbrian
Compton, 162 km
Lomonosov, 92 km

Late Imbrian
Cantor, 81 km

Copernican
Giordano Bruno, 22 km

The boundary where the flat floor of the proposed Near Side Megabasin meets the exposed slope of the basin is clearly visible near Fabry (184 km) and Compton (162 km). The Lomonosov-Fleming Basin obscures this edge below 30° south latitude. The Humboldtianum Basin can be seen in the upper left corner.

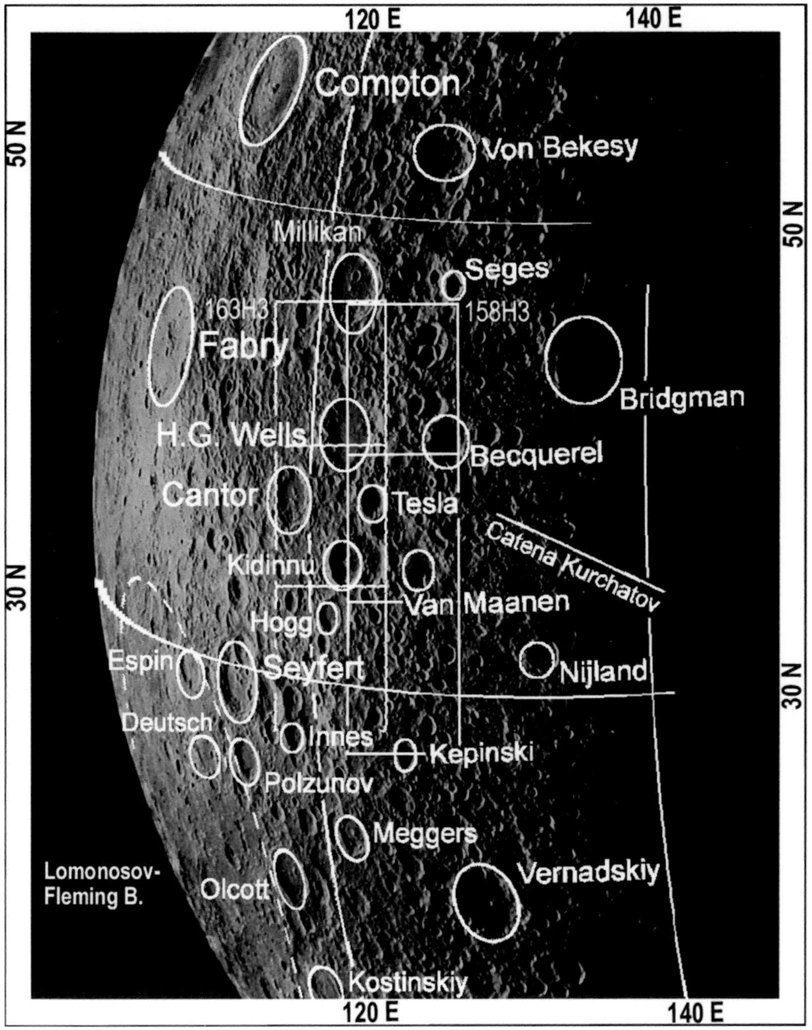

Distance: 1470km

Sun Elevation: 11°

LO5-158M and 163M

This is a mosaic of two medium-resolution frames with a large overlap. There are many chains of secondary craters here, such as Catena Kurchatov from the Humboldtianum Basin (to the northwest) and others from the Mendeleev Basin (to the southeast). The Lomonosove-Fleming Basin can also be seen in the photo on the preceding page.

The Northwestern Far Side Region: 97 The Moscoviense Basin

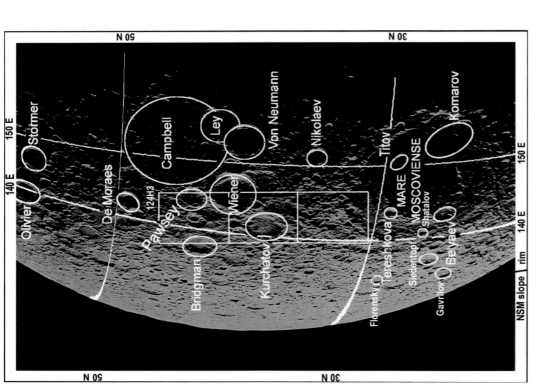

LO5-124M   Sun Elevation: 10.16°   Distance: 1478.87 km

The Moscoviense Basin and its ejecta blanket dominate this photo below 50° north latitude. The inner ring, filled with dark mare, is shown dramatically in this low sun angle image. The mare floor is clearer in the photo on the next page.

Key

Pre-Nectarian
Campbell, 219 km

Nectarian
Bridgman, 80 km
Moscoviense Basin, 445 km

Early Imbrian
Compton, 162 km

Late Imbrian
Nikolaev, 41 km
Størmer, 69 km
Von Neumann, 78 km

LO5-085M and 103M

Sun Elevation: 11°

Distance: 1515km

This image is a mosaic of two frames, each of which is included in the CD. This photo shows the detail within Mare Moscoviense. The dashed line shows the extreme boundary of that part of the ejecta from the proposed Near Side Megabasin that crossed over the megabasin's anti-node. Ejecta that had more velocity escaped the Moon.

Key

Pre-Nectarian
Campbell, 219 km

Nectarian
Appleton, 63 km
D'Alembert, 248 km
Freundlich, 85 km
Moscoviense Basin, 445 km
Wiener, 120 km

Early Imbrian
Compton, 162 km
Hutton, 50 km
Nijland, 35 km
Slipher, 69 km

Late Imbrian
Golovin, 37 km
Nikolaev, 41 km
Nusl, 61 km
Von Neumann, 78 km

Eratosthenian
Stearns, 36 km

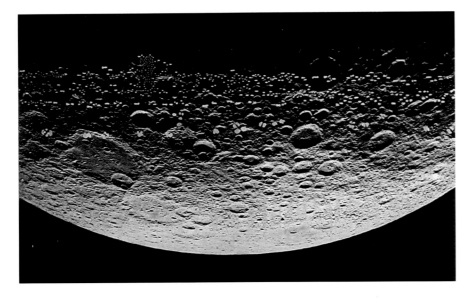

LO5-079M

Sun Elevation: 10.57°

Distance: 1529.62 km

The bubbles near the terminator are development arti-facts. The floor of D'Alembert (248 km) is well shown here, while the floor of Campbell is less clear; even the mare that is so striking in the image on the previous page is not vis-ible here.

Key

Pre-Nectarian
Campbell, 219 km
Freundlich-Sharonov
Basin, 600 km

Nectarian
Appleton, 63 km
D'Alembert, 248 km

Freundlich, 85 km
Larmor, 97 km

Early Imbrian
Hutton, 50 km
Slipher, 69 km

Late Imbrian
Golovin, 37 km
Nusl, 61 km
Von Neumann, 78 km

LO5-053M

Sun Elevation: 9.18°

Distance: 1449.77 km

This image spans the eastern and western hemispheres of the Moon. A large crater in the lower right corner is about the same size as Sommerfeld and Rowland but remains unnamed.

The Northwestern Far Side Region: **101** **The Moscoviense Basin**

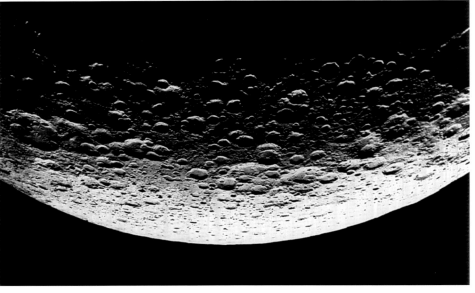

LO5-039M

Sun Elevation: 10.15°

Distance: 1571.18 km

This area is dominated by the ejecta from the pre-Nectarian Birkhoff Basin (Eastern Far Side Region). The floor of Rowland is clear in this image. The large unnamed crater south of Schneller (mentioned on the previous page) is also clearly shown here. Clementine data has revealed that Jackson has a large ray pattern.

Key

Pre-Nectarian
Cockcroft, 93 km
Evershed, 66 km
Mach, 180 km
Montgolfier, 88 km

Nectarian
Rowland, 171 km

Late Imbrian
Woltjer, 46 km

Copernican
Jackson, 71 km

From Compton

From Fabry

LO5-181H3   Sun Elevation: ~13°   Distance: ~1400 km

This is a region of the internal slope of the Near Side Megabasin. It has been covered with ejecta from the South Pole-Aitken Basin and then the Humboldtianum Basin but then the region in this subframe, at least its upper two-thirds, has been relatively free from the influence of other basins. The largest crater in this image is Petrie (**P**, Eratosthenian, 33 km).

Chains of secondary craters here come from Compton to the north and Fabry to the southwest. The low level of erosion of fine chains from Compton is evidence for assigning Compton's age to the Early Imbrian Period, after the heavy Nectarian bombardment.

S

Catena Sumner \

LO5-181H2   Sun Elevation: 13.32°   Distance: 1374.79 km

The north–northwest to south–southeast ridges and troughs running through this area are from the Humboldtianum Basin. Their patterns have obscured the crater Sisakyan (**S**, 34 km). The tip of Catena Sumner is near the bottom edge of this subframe (the catena is continued on the next page).

Catena Sumner

From Giordano Bruno

S

Sz

LO5-181H1   Sun Elevation: ~13°   Distance: ~1400 km

Catena Sumner is near the small, fresh crater Sumner (**S,** 50 km). The origin of Catena Sumner is not clear; it does not line up well with basins on the far side. It lines up best with the Imbrium Basin. Bright rays on the mare floor of Szilard (**Sz,** 120 km)) come from Giordano Bruno (Copernican, 22 km), which is out of the subframe to the west (see LO5-181M).

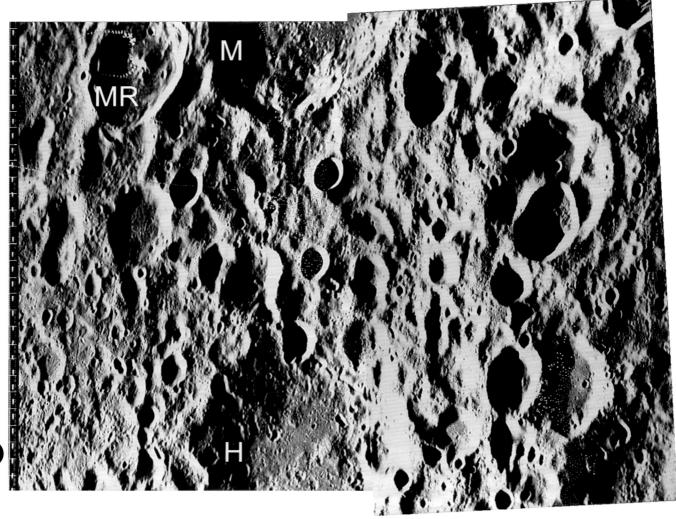

LO5-158H3 and -163H3   Sun Elevation: ~11°   Distance: ~1515 km

This region near Millikan (**M**, Nectarian, 98 km) contains the intersection of many influences. The ejecta blanket of the Humboldtianum Basin covers that of the Moscoviense Basin here (both are of the Nectarian Period as is Millikan). The younger crater just southwest of Millikan is Millikan R (**MR,** Early Imbrian, 49 km). H. G. Wells (**H,** 114 km) seems to have lived through all or many of these Nectarian events.

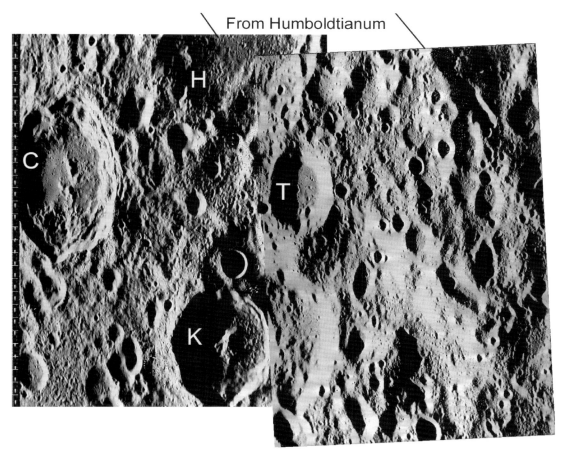

From Humboldtianum

LO5-158H2 and -163H2   Sun Elevation: ~11°   Distance: ~1515 km

The largest and oldest crater here is H. G. Wells (**H**, 114 km). Next in size, and much younger, is Cantor (**C**, Late Imbrian, 81 km). Kidinnu (**K**, Early Imbrian, 56 km) is the crater with the strong central peak, as is typical of craters of this size. The evidence for Kidinnu being older than Cantor is that ejecta from Cantor has modified the rim of Kidinnu. Tesla (**T**, 43 km) has retained a sharp rim, but its ejecta has been cratered nearly as much as that of Cantor, and is probably of the Nectarian Period.

This group of craters exemplifies the broad range of names approved through the nomenclature process of the International Astronomical Union. Wells (1886–1946) was a British science writer who authored "War of the Worlds." Georg Cantor (1845–1918) was born and died in Russia, but his mathematics work on set theory was in Germany. Moritz Cantor (1829–1920), a relative of Georg and a German mathematician, is also honored with this crater. Kidinnu (or Cidenas, died about 343 B.C.) was a Babylonian astronomer. Nikola Tesla (1856–1943) was a Croatian-born Serbian. He became an American inventor who influenced the growth of alternating current.

Ridges and deep troughs on the right side of this mosaic probably originate from the Humboldtianum Basin. In some places, the secondary craters are so close together that they appear as continuous ditches but some isolated craters are within the ditches. The ridges are raised by finer ejecta (note the pitting there), augmented by ejecta from the craters in the troughs.

From Mendeleev

LO5-158H1 and -163H1   Sun Elevation: ~11°   Distance: ~1515 km

Hogg (**H**, 38 km) is the largest distinct crater in this mosaic. Its impact has nearly hit the rim of the older, somewhat larger crater beneath it. This area is relatively free of influences from craters greater than 50 km in radius, and so the smaller craters have retained more sharpness than in other areas. The catenae in the right side of this image radiate from the Mendeleev Basin, although they have fallen on the exposed edge of the ejecta blanket of the Moscoviense Basin.

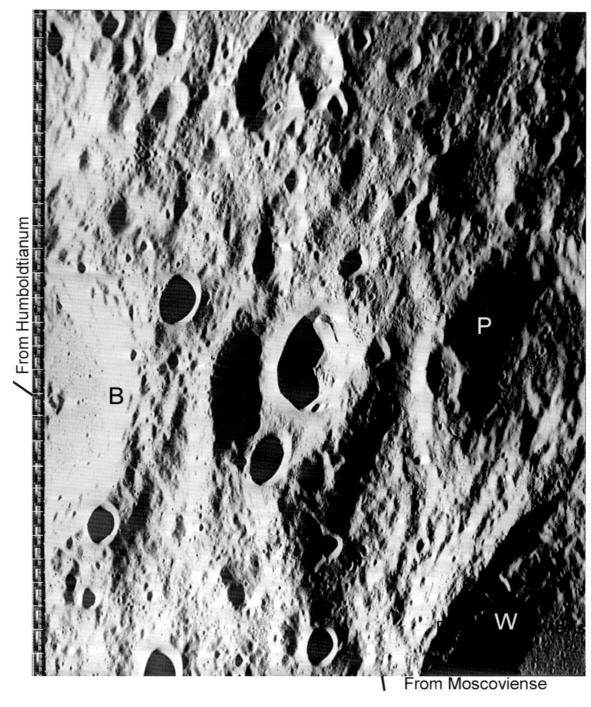

LO5-124H3   Sun Elevation: ~10°   Distance: ~1480 km

In this complex terrain of ridges and troughs, the outer boundary of the ejecta blanket of the Moscoviense Basin (to the south–southeast) lies under the ejecta blanket of the Humboldtianum Basin (to the northwest). Humboldtianum's ridges and troughs are exposed crossing Bridgman (**B**, Nectarian, 80 km) but stop east of Bridgman and in the vicinity of Pawsey (**P**, Nectarian, 60 km), leaving some of the larger striations from Moscoviense. Wiener (**W**, Nectarian, 120 km) came later, burying much of Pawsey.

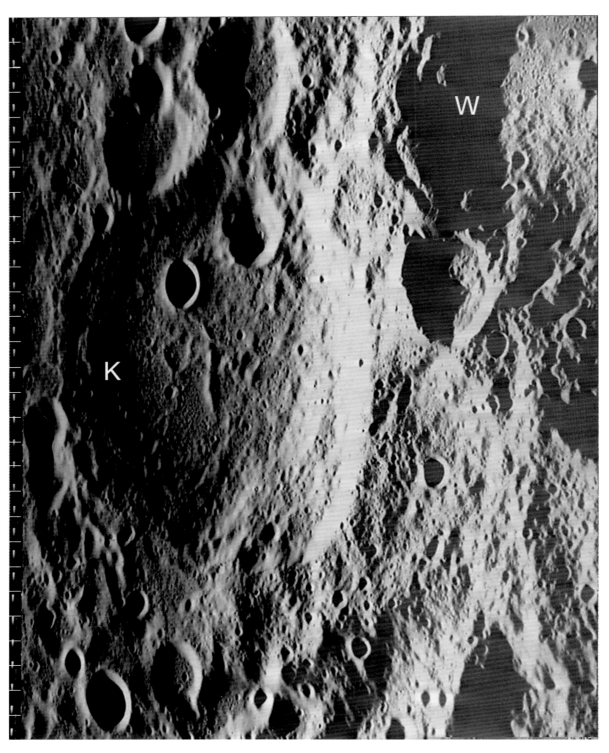

LO5-124H2   Sun Elevation: 10.16°   Distance: 1478.87 km

The large crater dominating this subframe is Kurchatov (**K**, 106 km). Both its internal depression and its external ejecta blanket have been covered with ejecta from Wiener (**W**, Nectarian, 120 km), making it difficult to determine whether Kurchatov is pre-Nectarian or Nectarian. The ejecta blanket of Kurchatov has resurfaced the patterns from earlier times, but a few earlier craters under that blanket are dimly revealed.

LO5-124H1   Sun Elevation: ~10°   Distance: ~1480 km

This is the northwest quadrant of the Moscoviense Basin (**M**, Nectarian, 445 km). Extensive cratering of the basin rim and ejecta blanket indicate that the impact that formed this basin happened quite early, but the surviving radial patterns in its ejecta suggest that it was not pre-Nectarian, and so it is assigned to the Nectarian Period. Within that period, overlapping ejecta patterns led the USGS geologists to place the Moscoviense Basin as older than the Mendeleev and Humboldtianum Basins. This subframe shows the rim of the Moscoviense basin, and below the rim is the depression, the top surface of the transient crater. For a high-resolution view of Mare Moscoviense and its contact with the depression surface, see the mosaic of LO5-085H1 and -103H1 on the next page.

Von Neumann

LO5-085H3 and LO5-103H3   Sun Elevation: ~11°   Distance: ~1515 km

Campbell (**C**, pre-Nectarian, 219 km) is the subject of this mosaic. Its original shape was circular (appearing elliptical in this oblique photo) but further impacts early in its history modified its rim. Specifically, Ley (**L**, 79 km) struck the inner edge of Campbell's southern rim and redistributed all of the rim material. Subsequently, in the Late Imbrian Period, Von Neumann (see next page) threw the southern rim of Ley onto its floor. Either one crater about the size of Ley or a cluster of smaller impacts split the eastern rim

(**Cr**) of Campbell into two ridges. Lava darkened by iron has found its way upward into the depression. This mare (**Cm**) is off-center because of diversion by the irregular rubble of ejecta that Von Neuman (next page) deposited on the floor of Campbell. This is an indication of the long time required for radioactivity to produce enough heat to melt the lava that rose into Campbell's floor. Campbell was formed in the pre-Nectarian Period, but the lava did not rise until the Late Imbrium Period.

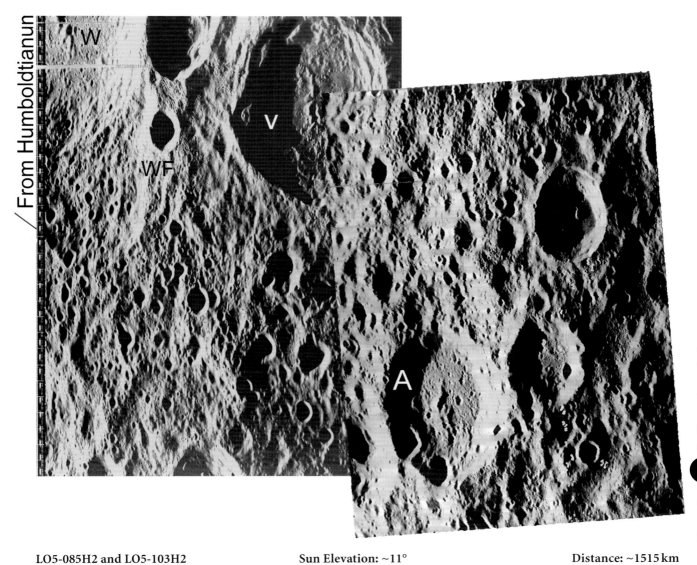

From Humboldtianun

W

V

WF

A

LO5-085H2 and LO5-103H2   **Sun Elevation: ~11°**   **Distance: ~1515 km**

The large crater near the top of this mosaic is Von Neumann (**V**, 78 km). Its floor is cluttered with ejecta from Appleton (**A**, 63 km), its neighbor to the south, and the young crater known as Wiener F (**WF**, 47 km) between Von Neumann and its large older neighbor Wiener (**W**, 120 km). The chains of craters here radiate not from Campbell but from the Humboldtianum Basin.

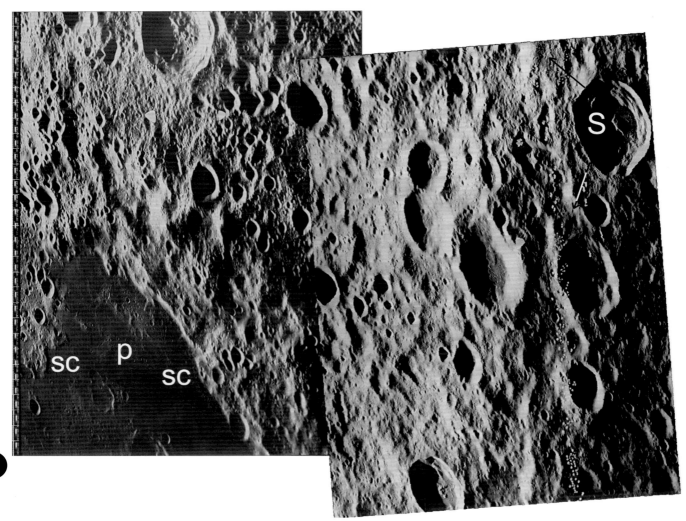

LO5-085H1 and LO5-103H1   Sun Elevation: ~11°   Distance: ~1515 km

The northeast quadrant of the Moscoviense Basin is shown in great detail in this mosaic. The lava flooding the floor rose higher than its current level, receding as it cooled and contracted. It has left a scarp (**sc**) about 20 m high, judging from the length of its shadow. This pattern is often seen at the edges of mare on the near side as well. The lava has covered plains areas (**p**) near the northern shore, darkening the surface, and then has flowed away, exposing the surface with some erosion. The rim of the Moscoviense Basin is rugged and rounded and, unlike many of the near side basins, lacks cliffs because the basin was formed in deep, fractured ejecta. Radial troughs and ridges can be seen in the ejecta blanket especially to the north, where the east-to-west lighting shows them clearly.

The impact of Stearns (**S**, 36 km) has been assigned by USGS to the Eratosthenian Period (the time after the Late Imbrian and before the Copernican Periods). The following passage, from "The Geologic History of the Moon" by Don Wilhelms, describes the reasons for the age assignment, based on the image above: freshness of Stearns' interior is evident, despite oblique viewing angle and observation by shadow. Smoothed radial ejecta and secondary craters are also preserved, despite rugged terra substrate; periphery of Stearns is pitted by small craters. No exterior textures would be visible around an Imbrian crater of same size and setting; fewer pits and sharper texture would characterize Copernican craters of same size and setting.

Rim of D'Alembert

L

C

LO5-079H3　　　　　　Sun Elevation: ~11°　　　　　　Distance: ~1530 km

The upper two-thirds of this subframe are in the ejecta blanket of D'Alembert (Nectarian, 246 km), whose rim runs in an arc just below the top of this subframe (see LO5-079M).

The blanket thins in an arc between Langevin (**L**, 58 km) and Chandler (**C**, 85 km).

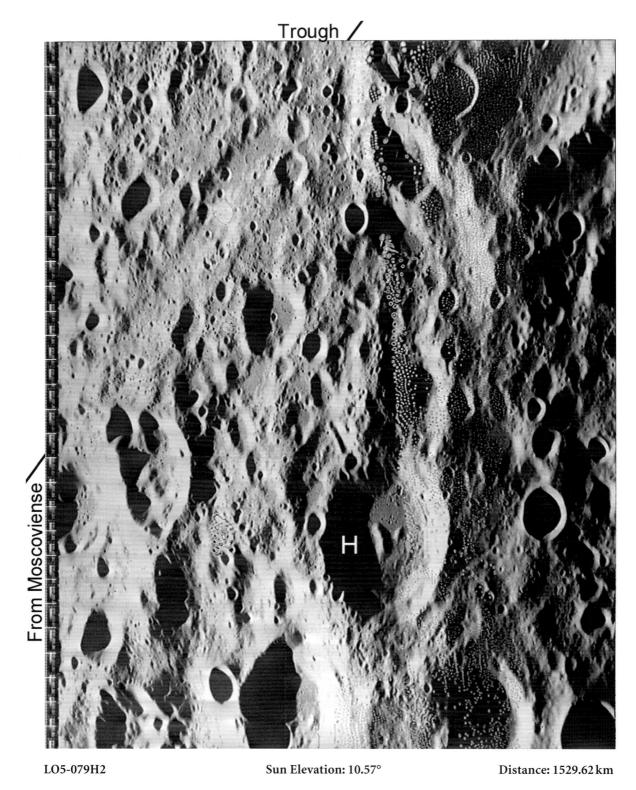

From Moscoviense

LO5-079H2   Sun Elevation: 10.57°   Distance: 1529.62 km

Hutton (**H**, 50 km) is a good example of a crater with a central peak, typical for its size. The smooth appearance of the material in the upper left may be due to finely divided ejecta.

A large trough with adjacent ridges runs across the upper left corner of this subframe. Both trough and fine ejecta are probably due to the Moscoviense Basin to the southwest.

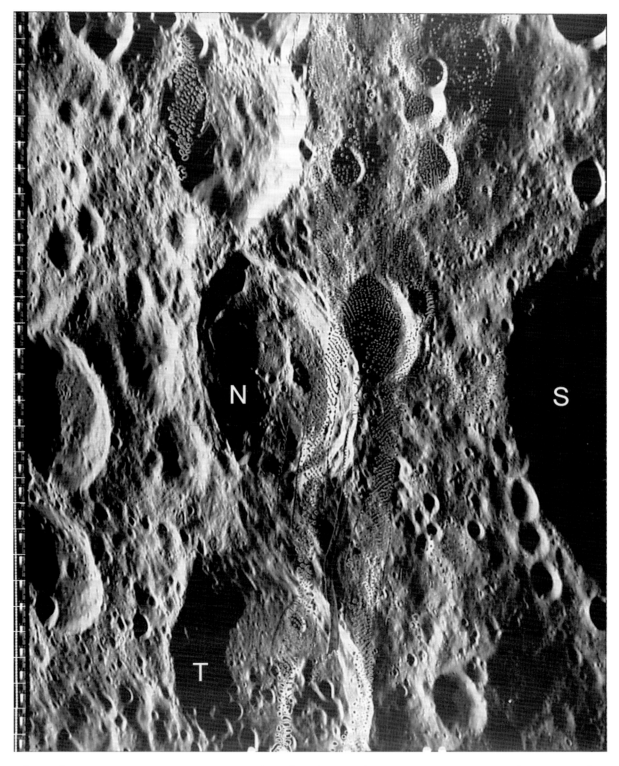

LO5-079H1   Sun Elevation: ~11°   Distance: 1530 km

Patterned groups of small craters, especially in lines like the four along the rim of Shayn (**S**, Nectarian or pre-Nectarian, 93 km) in the lower right corner, could be due to secondary impactors, although the source is unclear.

Nusl (**N**, Late Imbrian, 61 km) is relatively fresh, with sharp detail. It has thrown ejecta across the rim of Trumpler (**T**, Nectarian, 77 km).

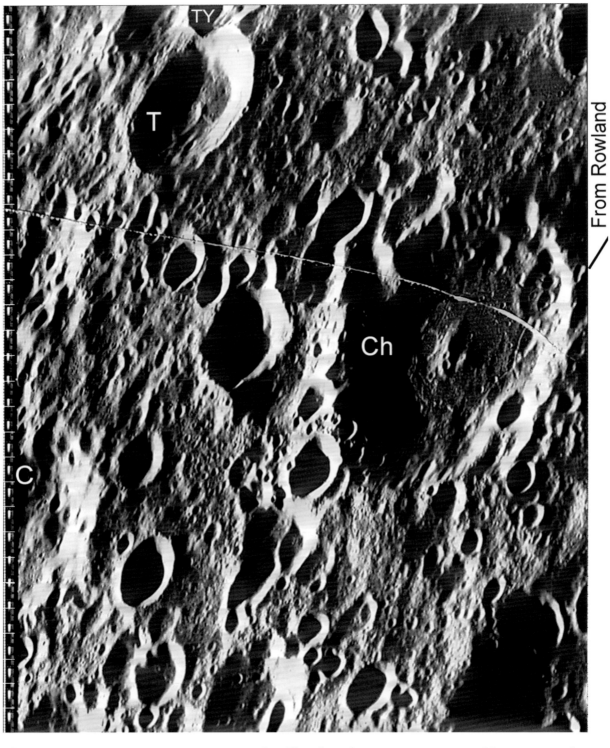

From Rowland

LO5-053H3   Sun Elevation: 9°   Distance: ~1450 km

The largest crater here is Chappell (**Ch**, pre-Nectarian, 80 km). The striations here come from Rowland (Nectarian, 171 km) to the northeast, whose ejecta blanket ends just beyond this subframe. Notice how these deposits have cut into the rim of Chappell and Cooper (**C**, pre-Nectarian, 38 km). Tsinger Y (**TY**, Eratosthenian, 31 km), on the northern rim of Tsinger (**T**, 44 km), is one of the very few young craters of its size in the Northwestern Far Side Region.

LO5-053H2   Sun Elevation: 9.08°   Distance: ~1449.77 km

The largest crater in this frame is Debye (**D**, pre-Nectarian, 142 km), somewhat more eroded than the younger craters that surround it. Perkin (**P**, Nectarian, 62 km) has impacted Debye's southern rim and thrown material from that rim onto the floor of Debye. Vertical scale can be estimated by the shadow that Perkin's rim throws on its floor. The shadow is 32 km long and the sun angle is 9°, and so the rim of Perkin is 5.4 km above the center of its floor. The material cast into Debye is about 20 km wide, 30 km long, and 1.5 km high, with a volume about 900 km³. The mass of such a volume is of the order of 2000×10^{12} kg (a million, million tons). All moved in less than a minute! In poetic justice, Perkin has had part of its southern rim dumped onto its floor by an even later impact.

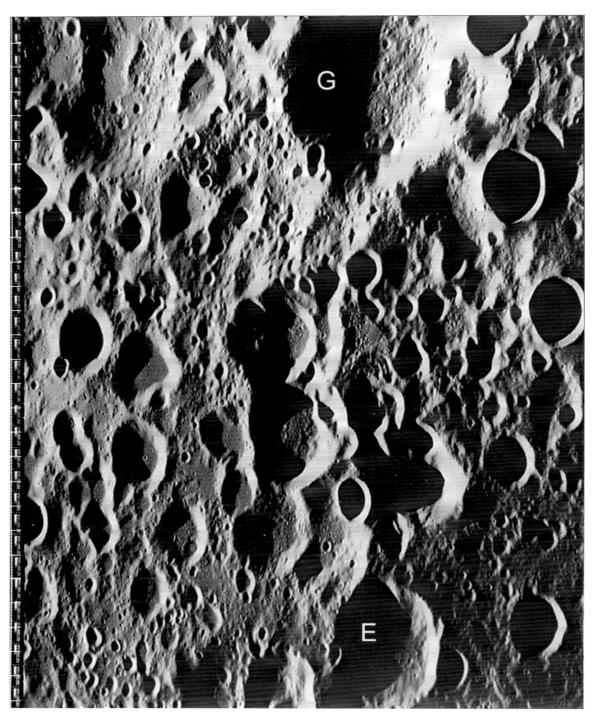

LO5-053H1   Sun Elevation: ~9°   Distance: ~1450 km

There is a cluster of craters here, each younger and smaller than Guillaume (**G**, Nectarian, 57 km) and Ehrlich (**E**, Nectarian, 30 km). Possibly their impactors were also clustered, fellow travelers from an impact or gravitational disruption. The extent of coverage of the floors of these craters illustrates that their ratios of depth to diameter increase as their size decreases. Nearly all craters less than 15 km in diameter have their floors shadowed by this sun elevation of 9°. Craters of this size that are not fully shadowed probably are tilted because they impacted into slopes.

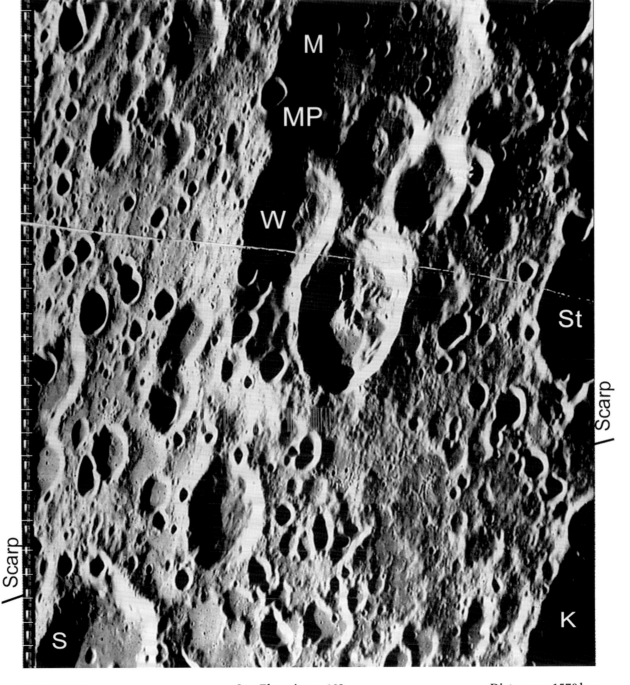

LO5-039H3   Sun Elevation: ~10°   Distance: ~1570 km

Ejecta from the Near Side Megabasin that passes over the antipode of that basin and returns to the Moon falls (according to the model) south of the arc labeled "Scarp." It passes north of the rim of Schneller (**S**, pre-Nectarian, 54 km) and between Kulik (**K**, 58 km) and Stoletov (**St**, Nectarian, 42 km). Unfortunately, the ejecta blankets of the Birkhoff Basin and of these nearby medium-sized craters obscure the difference in elevation, which may actually be spread out by variations in ejecta velocity. Woltjer (**W**, Late Imbrian, 46 km) and Montgolfier P (**MP**, Late Imbrian, 36 km) have impacted the earlier Montgolfier (**M**, pre-Nectarian, 88 km) crater.

LO5-039H2   Sun Elevation: 10.15°   Distance: 1571.18 km

This area illustrates the chaotic nature of random impacts and their effects. USGS characterizes this area as "Intercrater terrane." No heavy basin deposits have been detected here, and so the surface records the full history of impacts subsequent to that of the Near Side Megabasin and the South Pole-Aitken Basin. Schneller (**S**, pre-Nectarian, 54 km) is typical of the older craters. A pair of small young craters struck just north of Evershed (**E**, pre-Nectarian, 66 km) and drove material from Evershed's rim back onto its floor.

LO5-039H1   Sun Elevation: ~10°   Distance: ~1570 km

There appear to be several resurfaced patches on the floors of craters near Cockcroft (**C**, pre-Nectarian, 93 km) such as Cockcroft N (**CN**, 56 km), Van den Bergh (**V**, 42 km), and Van den Bergh Y (**VY**, 43 km). The plains northeast of Van den Bergh also appear to have been resurfaced. Clementine photos do not show a low albedo here, but the area is covered by rays from Jackson (see LO5-039M, Copernican, 71 km), which may obscure the albedo of mare or dark plains material.

Chapter 10

The Eastern Far Side Region: Birkhoff to Hertzsprung

10.1. Overview

This region extends from the Birkhoff Basin to the Hertzsprung Basin (see Figure 10.5 for detailed coverage).

The Eastern Far Side Region was photographed by Lunar Orbiter 5 in order to complete coverage of the far side. Because of mission constraints, the photographs were taken at an early, high orbit, which reduced the resolution. That is why this chapter covers a relatively large area with the usual number of photographs.

Remote sensing instruments have established that the area is unusually uniform in the distribution of chemical elements and minerals. Few areas of mare basalt have been identified. Apart from topography, the only major patterns visible to remote sensing are due to bright rays associated with young craters like Jackson and Ohm. The upper layers of the surface have been deposited by ejecta from a number of basins. As in the other far side regions, the lowest layers above the pristine crust were deposited by the proposed Near Side Megabasin and the South Pole-Aitcken Basin early in the pre-Nectarian Period, then came the Coulomb-Sarton Basin and the Birkhoff Basin later in the pre-Nectarian Period. The Korolev Basin came in the Nectarian Period and the Hertzsprung Basin came later in the Nectarian Period; its ejecta blanket overlays that of Korolev. Light ejecta came from the Imbrium Basin at the start of the Early Imbrian Period. Finally, the Orientale Basin covered most of this region at the end of the Early Imbrian Period.

10.2. The Birkhoff Basin

The Birkhoff Basin covered most of the northern part of this region with its ejecta at an early time. Yet an even earlier and larger feature, the Coulomb-Sarton Basin, underlies Birkhoff's ejecta.

The ancient floor of Birkhoff (Figure 10.1) records the subsequent history of several nearby craters: Sommerfield, Rowland, Stebbins, and Carnot.

Birkhoff is small for a basin, only about 10% larger in diameter than the conventional boundary between large craters and basins. There is a faint inner ring, 150 km in diameter. If there were outer rings, they have been obliterated by the surrounding large craters.

10.3. The Hertzsprung Basin

The Hertzsprung Basin is shown in Figure 10.2. Its ejecta once covered most of the central part of the Eastern Far Side Region. Orientale deposits have subsequently covered the eastern parts of the Hertzsprung Basin. Hertzsprung has a prominent inner ring that is 265 km in diameter.

Most of the chains of craters crossing the southeastern sector of Hertzsprung come from Orientale, but some are radial to the Mendel-Rydberg Basin, which is older than Orientale, but must be younger than Hertzsprung.

10.4. Rayed Craters

Figures 10.3 and 10.4 show the bright rays of Jackson and Ohm, as seen by Clementine at a wavelength of 415 nm. Jackson has been identified as Copernican because of its sharp features and relative lack of impacts in its internal cavity and external ejecta blanket.

Ohm was classified as Eratosthenian before this Clementine albedo data became available. Although they are somewhat dimmer than those of Jackson, the rays around Ohm are of the same composition as the underlying material and have not yet been fully darkened to maturity by the solar wind. This is evidence that Ohm is younger than is Copernicus (whose rays are nearly fully mature) and therefore its impact probably occurred in the Copernican period.

Robertson has a young topography but lacks a ray pattern. It was assigned to the Copernican Period but the lack of visible rays suggests that Robertson could be sufficiently older than Copernicus to be Eratosthenian.

10.5. Photos

Lunar Orbiter Medium-Resolution Frames

The approximate coverage of the medium resolution photos selected for the Eastern Far Side region is shown in Figure 10.5. All but one of these photos were taken by Lunar Orbiter 5 early in its mission (LO4-051 is the exception). Three tiers of pictures were taken; the northern tier near 60° north,

Figure 10.1. This high-resolution subframe shows the floor of the Birkhoff Basin (59° north, 147° west, pre-Nectarian, 330 km). The heavy bombardment by fairly large craters and the scattering of ejecta from nearby craters are evidence of its great age. LO5-029H2.

the central tier near 20° north and the southern tier near 20° south. Within each tier, the medium-resolution photos provide overlapping east–west coverage. The pictures were taken at high altitude and obliquely. Photos LO5-031M and 032M have been combined in a mosaic. LO5-025 M is not presented because it is redundant with LO5-029 and 008.

The medium-resolution photos are shown starting on the next page. They are shown in west-to-east order for each of the three tiers shown in Figure 10.5, starting with the northern tier and proceeding to the south. All of the photos presented after LO5-031 and 032 were taken at very long range (5000 to 6000 km from the targets) and extreme oblique angles. Consequently, the emphasis on annotating the named features was placed on the high-resolution subframes for these pictures.

All of the medium-resolution photos are shown before the high-resolution frames, to avoid breaking the continuity of contiguous coverage of the medium-resolution photos.

Lunar Orbiter High-Resolution Subframes

There are three high-resolution frames for each of the principal ground points labeled in Figure 10.5, except for LO4-051 and LO4-025. The H1, H2, and H3 high-resolution subframes of LO5-029, 025, and 006 overlap and are presented in the form of three mosaics; one for H3 (north), one for H2 (central), and one for H1 (south). The high-resolution subframes for LO5-031 and 032 overlap and all six subframes are combined in a single mosaic.

Figure 10.2. The Hertzsprung Basin (Nectarian, 1.5° north, 128.5° west, 540 km). Note the smaller density of large craters, in comparison with those in the floor of Birkhoff in Figure 10.1. LO5-024H1.

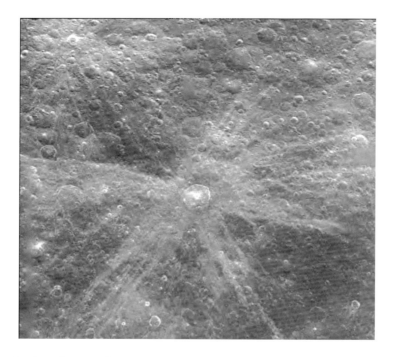

Figure 10.3. The ray pattern of Jackson (22.4° south, 163.1° west, Copernican, 71 km). NRL, brightness and contrast enhanced and filtered by this author.

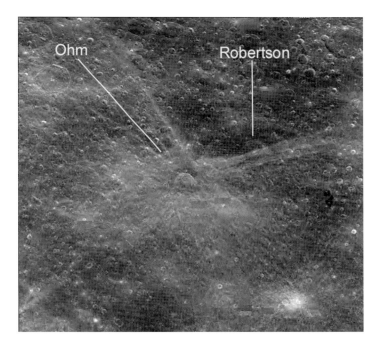

Figure 10.4. The ray pattern of Ohm (43.7° south, 113.5° W, Eratosthenian or Copernican, 64 km). Robertson (21.8° north, 105.2° west) is within the ray pattern of Ohm but lacks rays of its own. NRL, brightness and contrast enhanced and filtered by this author.

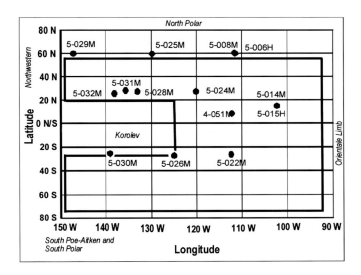

Figure 10.5. This figure labels the principal ground points and approximate total area of coverage of the medium-resolution Lunar Orbiter photos included in this chapter.

In this chapter, the high-resolution frame H2 is centered on the principal ground point, H1 is to the north of H2, and H3 is to the south of the central frame. The locations of the high-resolution subframes are shown precisely on the corresponding medium-resolution frames, except that high-resolution frame LO5-006H is marked on LO5-008M and LO5-015 on LO5-014M.

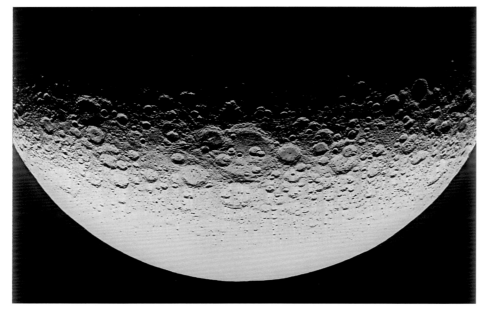

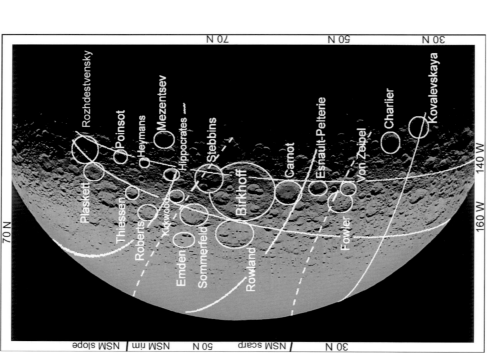

LO5-029M

Sun Elevation: 10.51°

Distance: 2673.36 km

Key

Pre-Nectarian
Birkhoff Basin, 330 km
Fowler, 146 km
Stebbins, 131 km

Nectarian
Carnot, PN?, 126 km
Rowland, 171 km
Sommerfeld, 169 km

Early Imbrian
Heymans, 50 km

Late Imbrian
Kovalevskaya, 115 km

Eratosthenian
Kirkwood, 67 km

The proposed Near Side Megabasin is so large that its rim extends over both poles. This and the photo on the next page show the northern rim of the Near Side Megabasin in the vicinity of Birkhoff. The sun angle is not favorable to show either rim or scarp here. The Carnot and Rowland events have followed the Birkhoff Basin event, but Carnot appears older and may have been in the pre-Nectarian Period.

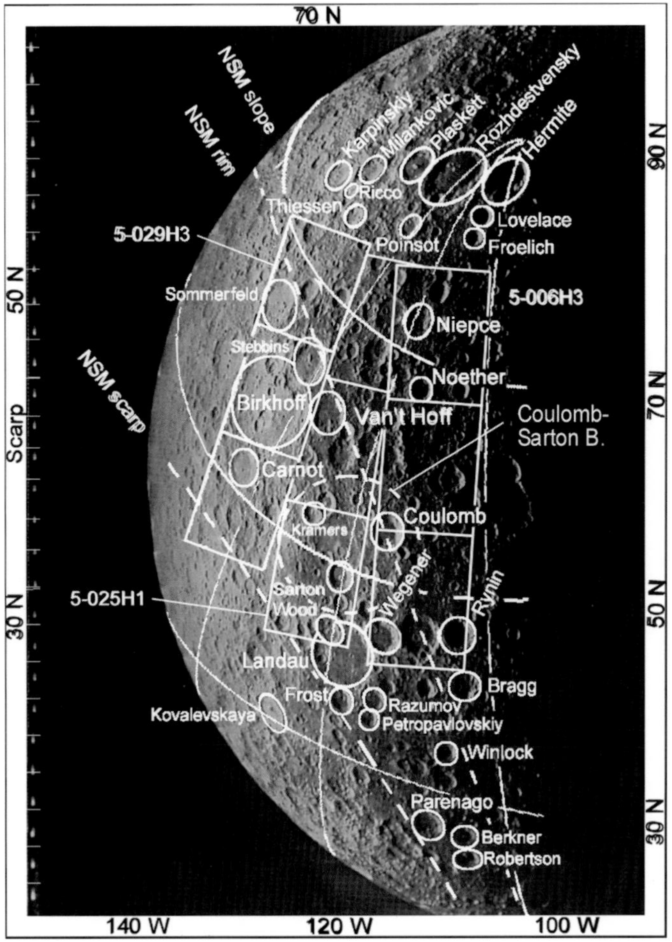

LO5-008M

Sun Elevation: 7.54°

Distance: 2753.15 km

East of the Birkhoff Basin is the even larger (and earlier) Coulomb-Sarton Basin. Ejecta from Birkhoff obscured the older basin. In the case of this photo, nearly all of the region of best coverage can be seen in high-resolution mosaics.

Key

Pre-Nectarian
Birkhoff Basin, 530 km
Coulomb-Sarton Basin, 530 km
Landau, 214 km
Stebbins, 131 km

Nectarian
Carnot, PN?, 128 km
Coulomb, PN?, 89 km
Sarton, 69 km
Sommerfeld, 169 km

Late Imbrian
Kovalevskaya, 115 km

Copernican
Robertson, 88 km

The Eastern Far **129** **Side Region**

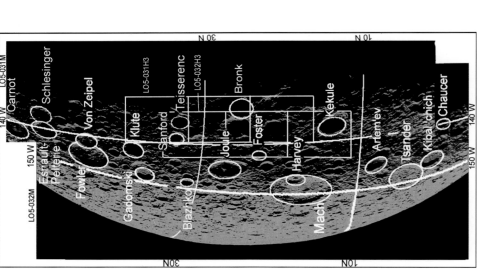

LO5-032M (left) and LO5-031M (right)   Sun Elevation: ~1.5°   Distance: ~1600 km

This area is northeast of Korolev and northwest of the Hertzsprung Basin. The ejecta blanket of Hertzsprung forms the upper layer of most of the area shown in this image. It overlays many of the features here, including the largest crater in this image, Mach.

Key

Pre-Nectarian
Fowler, 146 km
Gadomski, 65 km
Mach, 180 km
Schlesinger, 97 km
Tsander, 181 km

Nectarian
Carnot, PN?, 128 km
Joule, 96 km
Kibal'chich, 92 km

Early Imbrian
Artem'ev, 67 km

Late Imbrian
Blazhko, 54 km

Distance: 5265.72 km

Sun Elevation: 7.79°

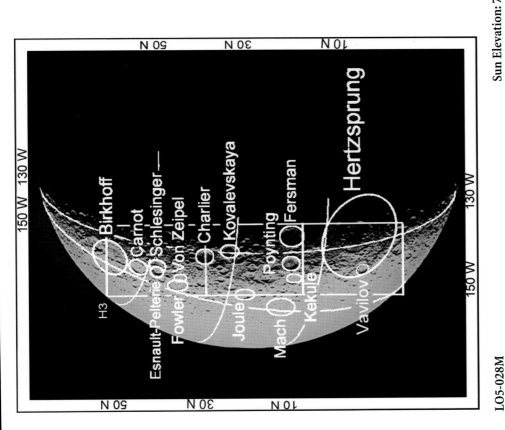

LO5-028M

This is the first of the photos taken from Lunar Orbiter 5 while it was still in a high orbit. The high-resolution subframes label additional features. The ejecta blanket of the Hertzsprung Basin constitutes the upper layer in most of this image but the blanket of the Orientale Basin (to the southeast) covers much of Hertzsprung itself, as well as Poynting and Fersman. Clementine brightness images show an extensive ray pattern associated with Vavilov.

Key

Pre-Nectarian
Fowler, 146 km
Mach, 180 km
Schlesinger, 97 km

Nectarian
Carnot, PN?, 128 km
Fersman, EI?, 151 km
Joule, 96 km
Poynting, 128 km
Hertzsprung Basin, 570 km

Late Imbrian
Kovalevskaya, 115 km

Copernican
Vavilov, 98 km

 The Eastern Far **131** Side Region

Distance: 5272.21 km

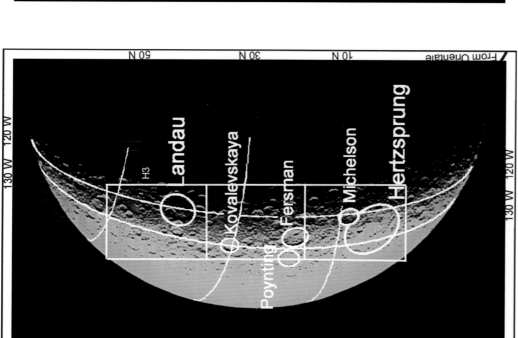

LO5-024M

Sun Elevation: 7.57°

Landau and the Coulomb-Sarton Basin have been covered by ejecta from the Birkhoff Basin. The eastern parts of those features have later been covered by ejecta from the Orientale Basin. There are linear features from the Orientale Basin near the terminator, like the chain of secondary craters in the southern part of the Hertzsprung Basin.

Key

Pre-Nectarian
Landau, 214 km

Nectarian
Fersman, El?, 151 km
Hertzsprung Basin, 570 km
Michelson, 123 km
Poynting, 128 km

Late Imbrian
Kovalevskaya, 115 km

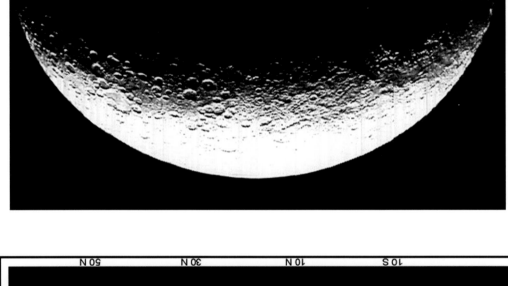

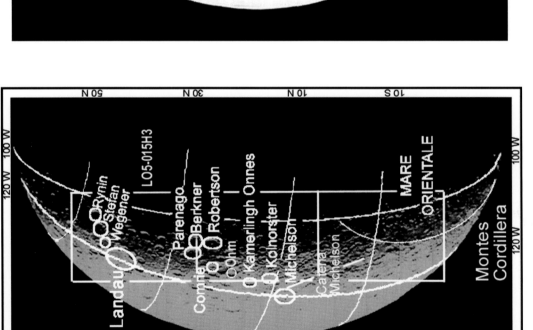

LO5-014M

Sun Elevation: 2.64°

Distance: 6000.34 km

The Orientale Basin is bounded by Montes Cordillera. Robertson (88 km) and Ohm are very young craters. See LO5-015H2 for a high-resolution view of them and a discussion of their ages.

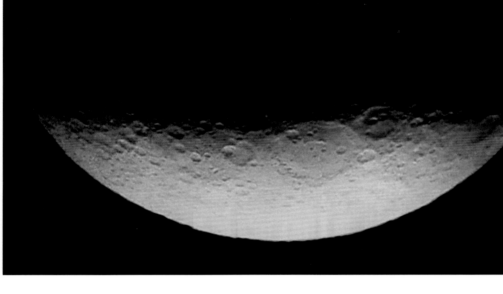

Distance: 6157.42 km

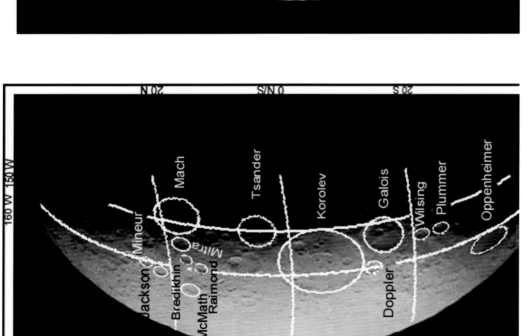

LO4-051M

Sun Elevation: N.A.

Key

Pre-Nectarian
Galois, 222 km
Mach, 180 km
Tsander, 181 km

Nectarian
Korolev Basin, 440 km
Oppenheimer, 208 km

Eratosthenian
Doppler, 110 km

Copernican
Jackson, 71 km

This photo, at the west edge of the southern tier of this region, shows major features surrounding the Korolev Basin. Doppler, Galois, and Oppenheimer are within the interior of the South Pole-Aitken Basin. Jackson, Mineur (73 km), and Mach are in the Eastern Far Side Region near the boundary of the Northwestern Far Side Region.

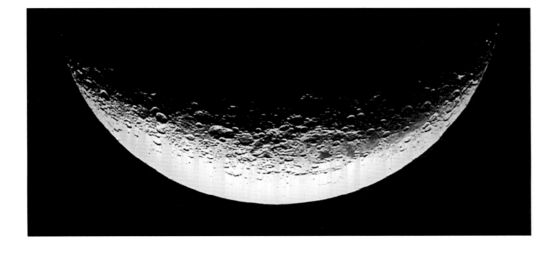

LO5-030M

Sun Elevation: 6.49°

Distance: 5391.42 km

The Apollo Basin is within the South Pole-Aitken Basin, as are Minkowski and Numerov. Galois, Mach, and Vavilov are in the Eastern Far Side Region. The ejecta blanket of the Hertzsprung Basin covers the northern half of this area. The western rim of Hertzsprung touches Vavilov.

Key

Pre-Nectarian
Apollo Basin, 505 km
Galois, 222 km
Mach, 180 km
Minkkowski, 113 km
Paschen, 124 km

Nectarian
Kibal'chich, 92 km
Numerov, 113 km

Copernican
Vavilov, 98 km

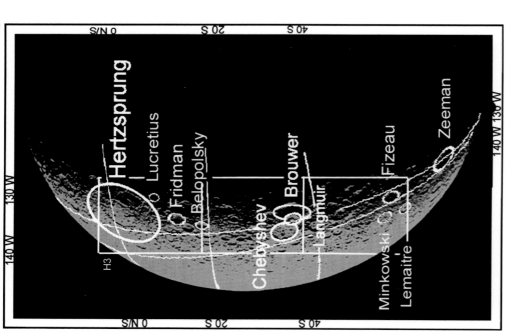

LO5-026M

Sun Elevation: 5.12°   Distance: 5380.83 km

The ejecta blanket of the Orientale Basin extends westward to about the 130° west meridian. This blanket overlies part of the Hertzsprung Basin. Fizeau and Fridman have deposited their ejecta over the Orientale blanket.

Key

Pre-Nectarian
Minkowski, 113 km

Nectarian
Brouwer, El?, 158 km
Chebyshev, 178 km
Hertzsprung Basin, 570 km
Zeeman, 190 km

Late Imbrian
Fizeau, 111 km
Fridman (Friedmann), 102 km

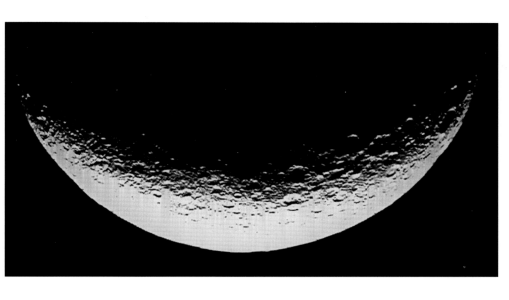

LO5-022M

Sun Elevation: 5.34°

Distance: 5441.61 km

The terminator in this photo is near 110° west longitude, so it is quite close to the limb. Montes Cordillera is the rim of the Orientale Basin. The Orientale Basin and the Imbrium Basin had all or parts of their circular rims named as mountain ranges before they were recognized as the rims of basins. Many ridges, troughs, and crater chains such as Catena Michelson radiate from Orientale.

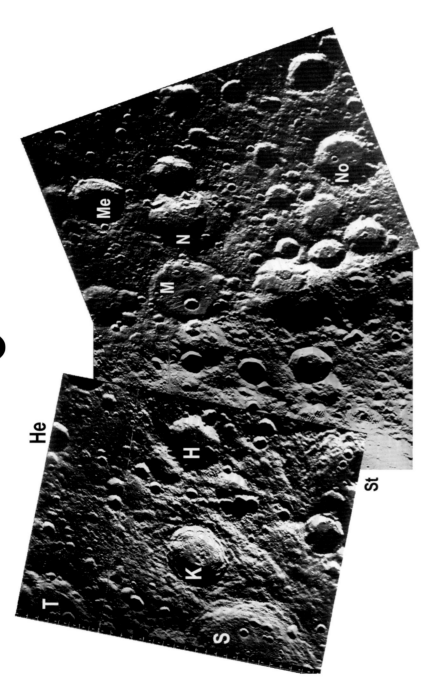

LO5-029H3 (left), 025H3 (center), and LO5-006H3 (right)

Sun Elevation: ~8°   Distance: ~5270 km

See LO5-008 for the location of these subframes. The ejecta blanket of
the Birkhoff Basin covers this area. Later craters such as Sommerfeld,
Kirkwood, and Hipppocrates have modified this ejecta blanket, without
changing its character. They have added their crater chains in an inter-
laced pattern, including contributions from Rowland and Stebbins.

Key

Pre-Nectarian
Stebbins, **St**, 131 km

Nectarian
Sommerfeld, **S**, 169 km

Early Imbrian
Heymans, **He**, 50 km

Unassigned
Hippocrates, **H**, 60 km
Kirkwood, **K**, 67 km

Merrill, **Me,** 57 km
Mezentsev (**M**, 89 km)
Niepce (**N**, 57 km)
Noether (**No**, 67 km)
Thiessen (**T**, 66 km)

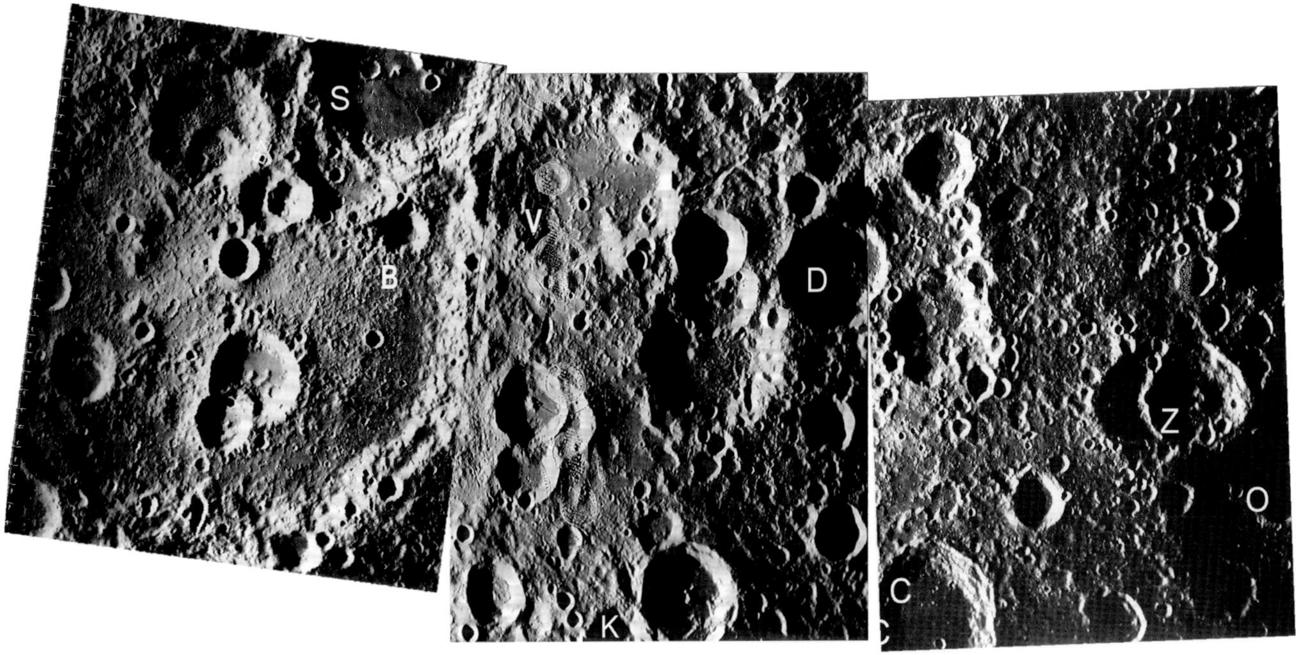

LO5-029H2 (left), 025H2 (center), and LO5-006H2 (right)

Sun Elevation: ~8°   Distance: ~5270 km

This mosaic shows the Birkhoff Basin and its eastern ejecta. The great age of this feature is established by the large numbers of craters of considerable size within the basin and in its ejecta blanket. The floor of Stebbins has been resurfaced. If it is mare material, the viscosity must have been higher than is usual. The striations between Dyson and Zsigmondy may be coming from Sommerfeld.

Key

Pre-Nectarian
Stebbins, **S** 131 km
Birkhoff Basin, **B**, 330 km

Nectarian
Coulomb, **C**, PN?, 89 km

Unassigned
Dyson, **D**, 63 km
Kramers, **K**, 81 km
Omar Khayyam, **O**, 70 km
Van't Hoff, **V**, 92 km
Zsigmondy, **Z**, 65 km

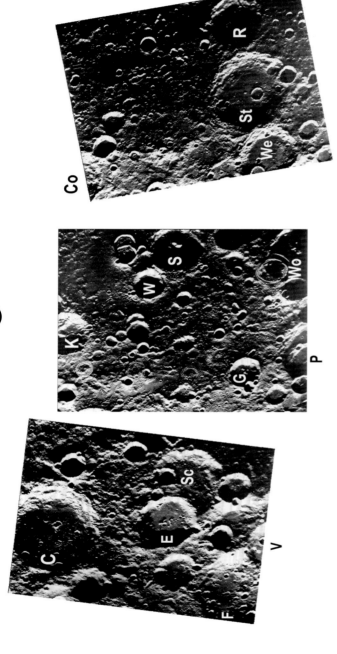

LO5-029H1 (left), 025H1 (center), and LO5-006H1 (right)

Sun Elevation: ~8°

Distance: ~5270 km

There are indications that mare has invaded the floor of Carnot, especially in its northeast sector. Carnot has succeeded not only Birkhoff (preceding page), but also Esnault-Pelterie, which might also be even older than Birkhoff. Fowler and Schlesinger came before Birkhoff and carry the marks of its ejecta and possibly Carnot's as well. Esnault-Pelterie followed Birkhoff, but later received ejecta from Carnot. An area of dark plains has formed northeast of Weber.

Key

Pre-Nectarian
Fowler, **F**, 146 km
Schlesinger, **Sc**, 97 km

Nectarian
Carnot, **C**, 126 km
Coulomb, **Co**, PN?, 89 km

Esnault-Pelterie, **E**, **PN?**, 79 km
Sarton (**S**, 69 km)

Early Imbrian
Gullstrand, **G**, 43 km

Unassigned
Kramers, **K**, 81 km

Perrine, **P**, 86 km
Rynin, **R**, 75 km
Stefan, **St**, 125 km
Von Zeipel, **V**, 83 km
Weber, **W**, 42 km
Wegener, **We**, 88 km
Wood, **Wo**, 78 km

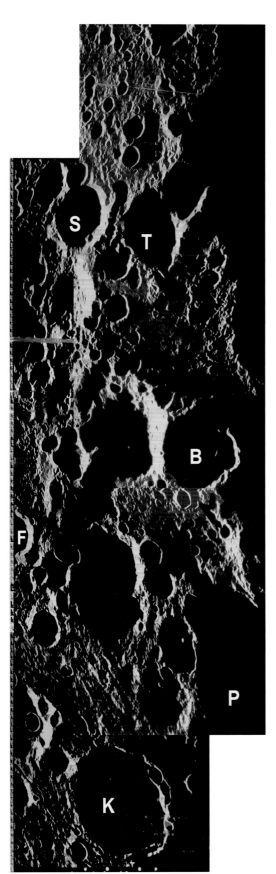

LO5-032H (Left) and LO5-031H (Right)
Sun Elevation: ~1.5°;
Distance: ~1600 km

This mosaic is composed of subframes whose information content has suffered from extremely low sun elevation. The floors of craters are fully shadowed and some of the rims throw shadows on their ejecta blankets. On the other hand, the photo is valuable because the texture of relatively flat areas between craters is emphasized. This area is within the ejecta blanket of Hertzsprung and the northern part of the image has received an additional layer of ejecta from Orientale.

This region appears in better light in subframe LO28 H2 on the page after next.

Key

Pre-Nectarian
Fowler, **F**, 146 km

Nectarian
Poynting, **P**, 128 km

Unassigned
Bronk, **B**, 64 km
Kekule, **K**, 94 km
Sanford, **S**, 55 km
Teisserenc, **T**, 62 km

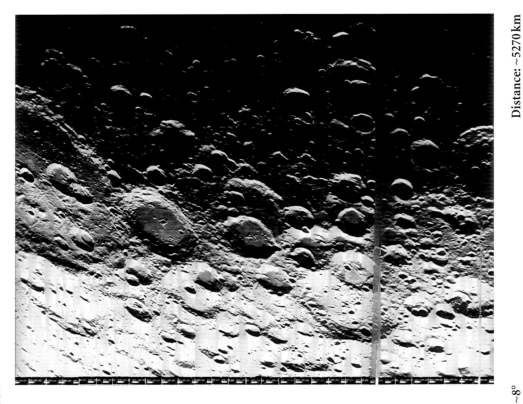

LO5-028H3

Sun Elevation: ~8°

Distance: ~5270 km

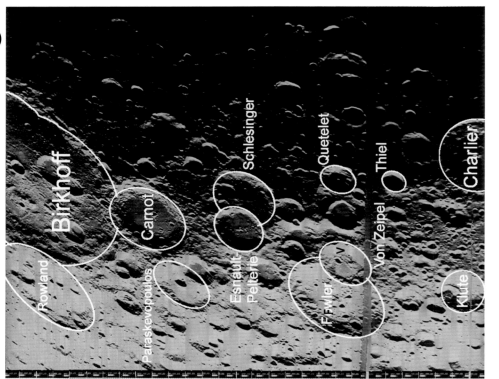

This view of the Birkhoff Basin from the south shows how Rowland and Carnot modified its rim. The crater chains that cross Rowland, Carnot, Esnault-Pelterie, and Schlesinger appear to come from D'Alembert to the west, establishing D'Alemebert as the youngest of these Nectarian craters.

Key

Pre-Nectarian
Birkhoff Basin, 330 km
Charlie, 99 km
Fowler, 146 km
Schlesinger, 97 km
Von Zeipel, 83 km

Nectarian
Carnot, 126 km
Rowland, 171 km
Esnault-Pelterie, PN?, 79 km

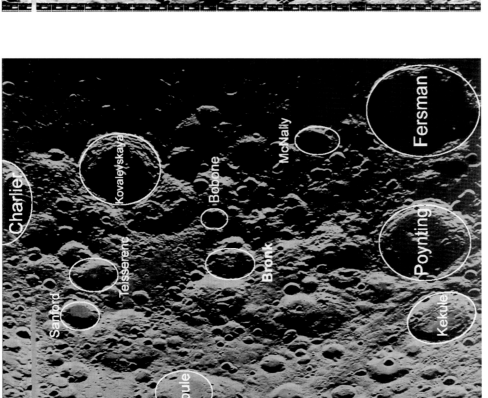

LO5-028H2

Sun Elevation: 7.79°

Distance: 5265.72 km

Most of this area contains overlapping pre-Nectarian craters like Charlier. Poynting came later and Fersman came after Poynting. Kovalevskaya was the last to arrive on this scene and is nearly unmarked. A tongue of Orientlale ejecta reaches out between Bronk and Kovaleskaya.

Key

Pre-Mectarian
Charlier, 99 km

Nectarian
Fersman, El?, 53 km
Joule, 96 km
Poynting, 128 km

Late Imbrian
Kovalevskaya, 115 km

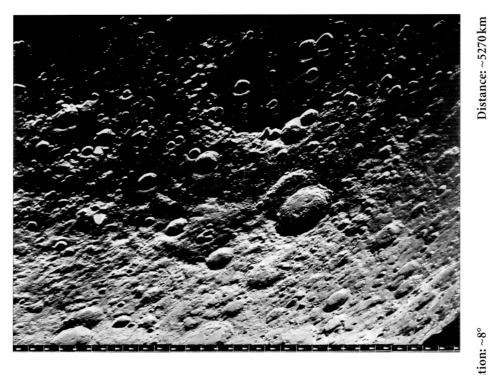

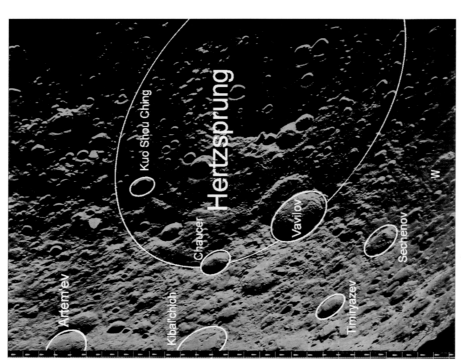

LO5-028H1

Sun Elevation: ~8°

Distance: ~5270 km

Vavilov, on the main ring of the Hertzsprung Basin, is nearly free of later impacts. It also has a ray pattern. The inner ring of Hertzsprung (265 km) is shown clearly here, as is an intermediate ring (410 km) that passes under Kuo Shou Ching (34 km).

Key

Pre-Mectarian
Hertzsprung Basin, 570 km

Nectarian
Kibal'chich, 92 km
Sechenov, 62 km
Timiryazev, 53 km

Early Imbrian
Artem'ev, 67 km

Copernican
Vavilov, 98 km

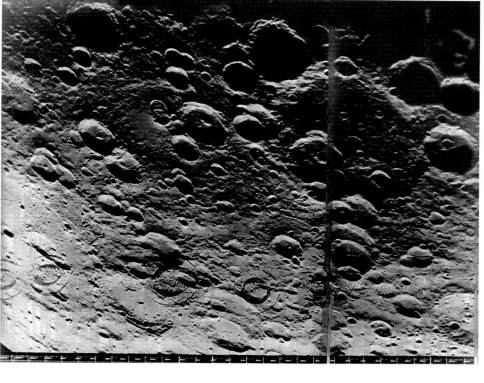

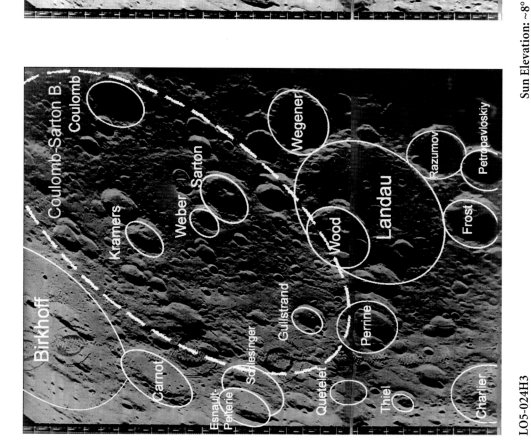

Sun Elevation: ~8°

Distance: ~5270 km

LO5-024H3

Key

Pre-Nectarian
Birkhoff Basin, 330 km
Charlier, 99 km
Coulomb-Sarton Basin, 630 km
Landau, 214 km

Nectarian
Carnot, 126 km
Coulomb, 89 km
Sarton, 69 km
Schlesinger, 97 km

Early Imbrian
Gullstrand, 43 km
Thiel, 32 km

The characteristics of large craters are similar to those of small basins. This image includes both the Birkhoff Basin and the large crater Landau. They have similar ejecta blankets, each with a typical pattern of radial troughs and ridges but only Birkhoff has an inner ring. The Coulomb-Sarton Basin is between Birkhoff and Landau, formed at an even earlier time. Kramers and Sarton are at its inner ring. The craters that have been used to name the Coulomb-Sarton Basin are related only by position.

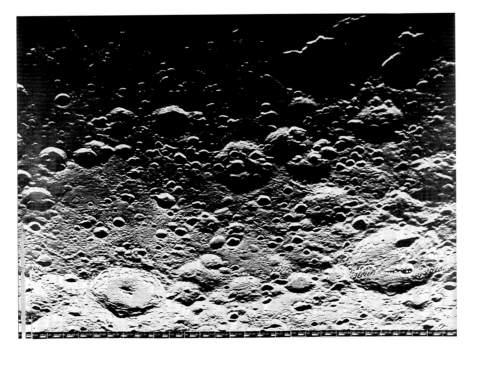

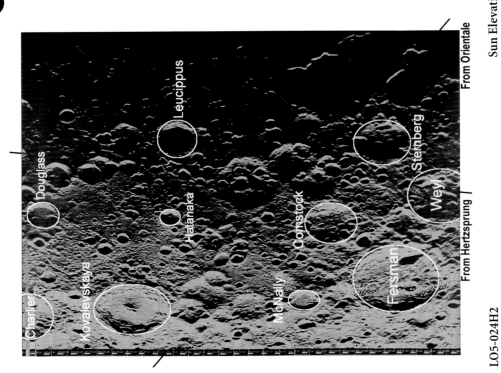

From Hertzsprung

From Orientale

Charlier

Kovalevskaya

Douglass

Leucippus

Hatanaka

McNally

Comstock

Fersman

Weyl

Steinberg

LO5-024H2

Sun Elevation: 7.57°

Distance: 5272.21 km

The northern half of this subframe exhibits ejecta, including chains of craters, from the Hertzsprung Basin. In the southern half, ejecta from the younger Orientale Basin dominate. Because of this double bombardment, age assignments of many craters in this region and to the south are uncertain.

Key

Pre-Nectarian
Charlier, 99 km
Weyl, N?, 108 km

Nectarian
Fersman, LI?, 151 km

Late Imbrian
Kovalevskaya, 115 km

LO5-024H1

Sun Elevation: ~8°

Distance: ~5270 km

This is a view of the eastern portion of the Hertzsprung Basin, showing the structure of its main ring (570 km) and inner ring (265 km). There are two fainter and discontinuous internal rings at diameters of 140 km and 410 km. Orientale sent sprays of secondary impactors that form chains of craters, including Catena Michelson, and Catena Lucretius in this area. However, the north-central region of Hertzsprung was either spared or resurfaced with mare.

Key

Nectarian
Hertzberg Basin, 570 km
Michelson, 123 km

Early Imbrian
Grigg, 36 km

Late Imbrian
Lucretius, 63 km

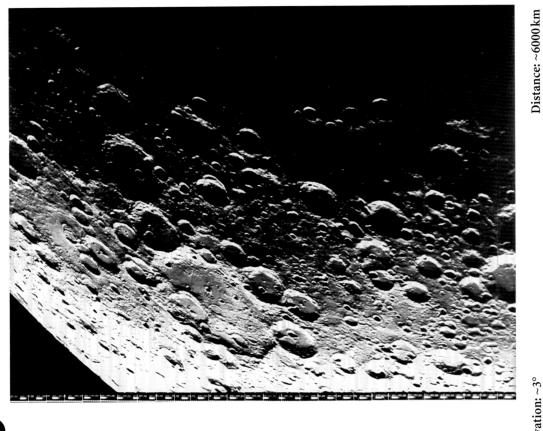

LO5-015H3

Sun Elevation: ~3°

Distance: ~6000 km

The Lorentz Basin is older than Birkhoff but younger than the Coulomb-Sarton Basin. Lorentz extends all the way to the eastern limb of the far side (the western limb of the near side). Northwest of Winlock, the ejecta blanket of Lorentz dominates, but to the southwest it is covered with ejecta from the Orientale blanket.

Key

Pre-Nectarian
Coulomb-Sarton Basin
Landau, 214 km
Lorentz Basin, 360 km

Nectarian
Coulomb, El?, 89 km

Early Imbrian
Gullstrand, 43 km

Late Imbrian
Lacchini, 58 km

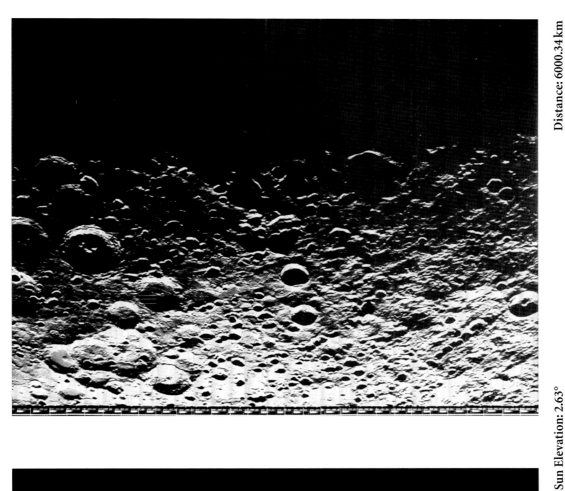

LO5-015H2

Robertson and Ohm are young, sharply defined craters. Robertson has been classified as Copernican and Ohm as Eratosthenian but Clementine brightness data shows that Ohm has a strong ray pattern and Robertson has none. This is evidence for reversing the classifications. This area is close to the Orientale Basin (next page) and is covered with its ejecta blanket. Catena Leuschner, radial to Orientale, takes its name from the unrelated crater Leuschner.

Sun Elevation: 2.63°

Distance: 6000.34km

Key

Late Imbrian
Butlerov, 40 km
Pease, 38 km

Erastothenian
Leuschner, 49 km
Ohm, 64 km

Copernican
Robertson, 88 km

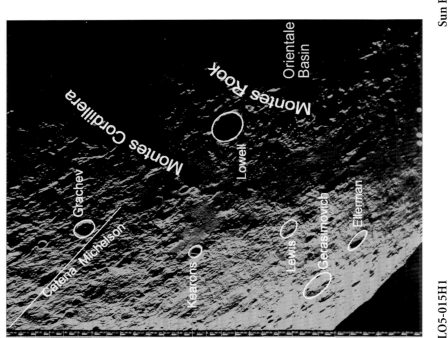

LO5-015H1

Sun Elevation: ~3°

Distance: ~6000 km

Key

Early Imbrian
Montes Cordillera, 930 km
Montes Rook, 620/480 km
Orientale Basin, 930 km

Late Imbrian
Kearons, 23 km
Lowell, 66 km

Eratosthenian
Grachev, 43 km

At last we encounter the Orientale Basin, the source of the ejecta blanket, crater chains, and other secondary craters on previous pages. The western ejecta blanket is revealed in great detail here. The rim of this basin is named Montes Cordillera. Inside the rim, Orientale has a pair of rings, the smaller the inner ring and the larger the intermediate ring. Montes Rook is the name of the complex composed of these two rings. A fuller presentation of this basin is in the Orientale Limb Region chapter.

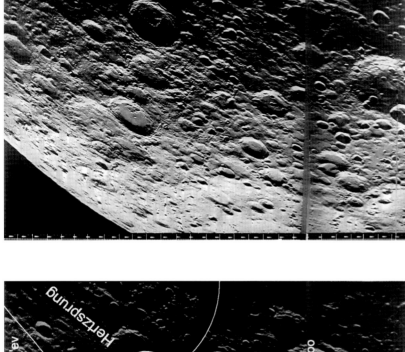

LO5-030H3

Sun Elevation: ~6.5°

Distance: ~5390 km

This image includes part of the Korolev Region. The ejecta blanket of the Hertzsprung Basin dominates this subframe; Vavilov is mostly inside the rim of that basin, and Chaucer is outside of the rim. Although Korolev and Hertzsprung are both Nectarian, it is clear that the Korolev impact was earlier in that period. Paschen and Tsander have been buried by both Korolev and Hertzsprung.

Key

Pre-Nectarian
Galois, 222 km
Paschen, 133 km
Tsander, 181 km

Nectarian
Chaucer, 45 km
Hertzsprung Basin, 670 km
Kibal'chich, 92 km
Korolev Basin, 440 km
Sechenov, 62 km
Timiryazev, 53 km

Early Imbrian
Artem'ev, 67 km

Late Imbrian
Dirichlet, 47 km

Copernican
Vavilov, 98 km

Distance: 5391.42 km

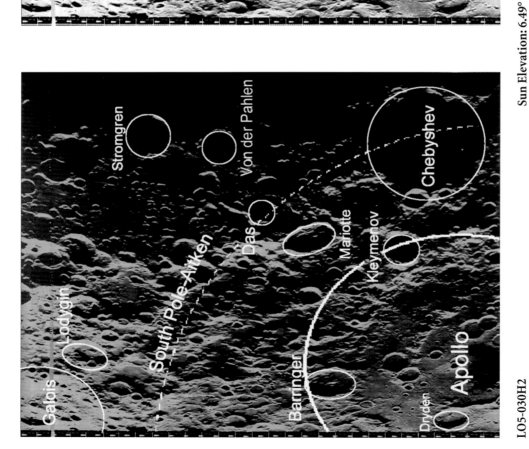

LO5-030H2

Sun Elevation: 6.49°

Mariotte has double boundary roles. It is at the edge of the ejecta blanket of the Hertzsprung Basin and it is on the rim of the South Pole-Aitken Basin. Kleymanov is on the rim of the Apollo Basin; Das has a significant ray pattern that is not visible here, but is revealed by Clementine data.

Key

Pre-Nectarian
Apollo Basin, 505 km
Galois 222 km
South Pole-Aitken Basin, 2500 km

Nectarian
Chebyshev, PN?, 1778 km

Late Imbrian
Mariotte, 65 km

Copernican
Das, 38 km

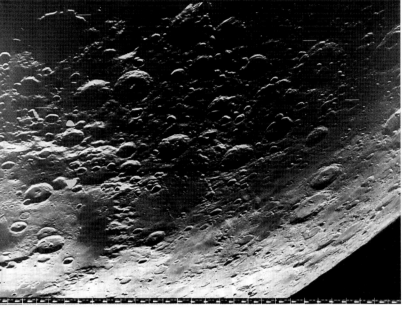

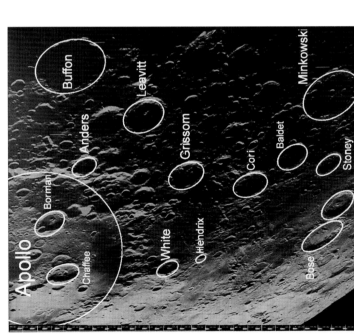

LO5-030H1

Sun Elevation: ~6.5°

Distance: ~5390 km

The area of this entire subframe is within the South Pole-Aitken Basin. Buffon is on that basin's rim. The crater chains crossing the resurfaced mare area between Cori and Hendrix are radial to the Orientale Basin, to the northeast. If they are from that basin, the last lava flow here was in the Early Imbrian Period at the latest.

Key

Pre-Nectarian
Apollo Basin, 505 km
Minkowski, 113 km

Nectarian
Bhabha, 64 km
Bose, 91 km
Buffon, 106 km

Early Imbrian
Borman, 50 km

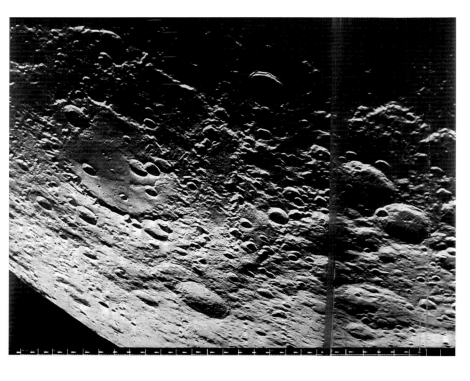

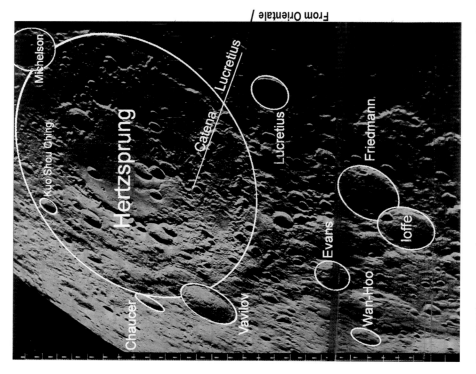

From Orientale /

LO5-026H3

Sun Elevation: ~5°

Distance: ~5380km

This view of the Hertzsprung Basin from the south shows excellent definition of its ring structure. The inner and intermediate rings are particularly well defined in this image. Catena Lucretius is formed by aligned secondary impactors from the Orientale Basin and is representative of many such crater chains.

Key

Pre-Nectarian
Hertzsprung Basin, 570 km

Nectarian
Chaucer, 45 km
Michelson, 123 km

Late Imbrian
Kuo Shou Ching, 34 km
Lucretius, 63 km
Fridman (Friedman), 102 km

Copernican
Vavilov, 98 km

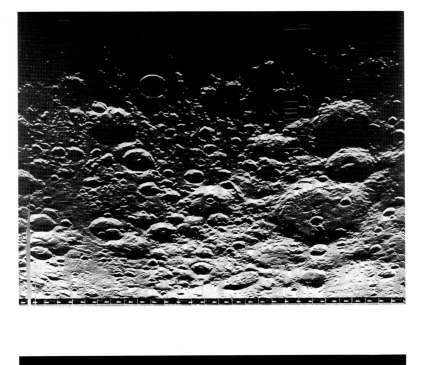

Belopol'skiy

Gerasimovich

Ellerman

Strømgren

Von der Pahlen

Das

Mariotte

Chebyshev

Brouwer

Langmuir

Kleymenov

South Pole-Aitken Basin

South Pole-Aitken Basin

LO5-026H2

Sun Elevation: 5.12°

Distance: 5380.83 km

This area, south of the Hertzsprung Basin and east of the ApolloBasin, has been subject to several major influences. It covers the northeast sector of the South Pole-Aitken Basin and craters such as Chebyshev and Brouwer are on or near that basin's rim. Then Hertzsprung threw its ejecta as far as Chebyshev. Later, the Orientale Basin's ejecta also reached Chebyshev.

Key

Nectarian
Brouwer, PN? 158 km
Chebyshev, PN? 178 km

Late Imbrian
Mariotte, 65 km

Copernican
Das, 38 km

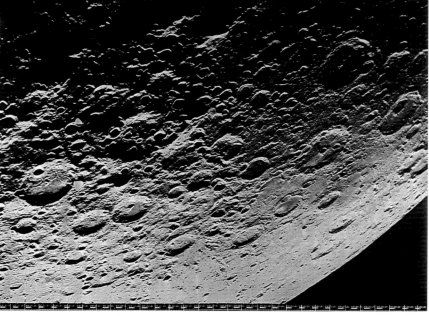

From Mendel-Rydberg

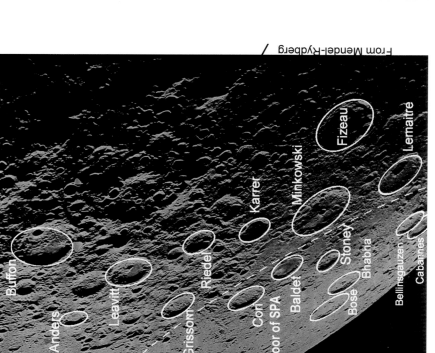

LO5-026H1

Sun Elevation: ~5°

Distance: ~5380 km

Key

Pre-Nectarian
Minkowski, 113 km

Nectarian
Bhaba, 64 km
Bose, 91 km
Buffon, 106 km
Lemaitre, 91 km

Late Imbrian
Fizeau, 111 km

The area within the dashed elipse is the flat floor of the South Pole-Aitken Basin and the area outside of the line is the slope rising to the rim. This area has clearly been resurfaced, perhaps in the pre-Nectarian Period. Both Lemaitre and Fizeau are Nectarian, but Fizeau is younger; ejecta from Fizeau is on the ejecta blanket of Lemaitre. Striations from the southeast running north of Fizeau are from the Mendel-Rydberg Basiin, all the way from the limb.

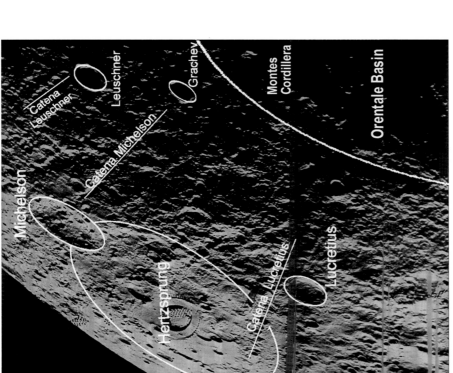

LO5-022H3

Sun Elevation: 5°

Distance: ~5440 km

This subframe shows the area between the Hertzsprung Basin and the Orientale Basin. Since Orientale came later, its ejecta pattern dominates the area. Three named crater chains Catena Lucretius, Catena Michelson, and Catena Leuschner illustrate how the multitude of striations here converge on their sources in the Orientale Basin. Montes Cordillera continues on the next page.

Key

Nectarian
Hertzsprung Basin, 570 km
Michelson, 123 km

Early Imbrian
Montes Cordillera, 930 km
Orientale Basin, 930 km

Late Imbrian
Lucretius, 63 km

Eratosthenian
Grachev, 35 km
Leuschner, 49 km

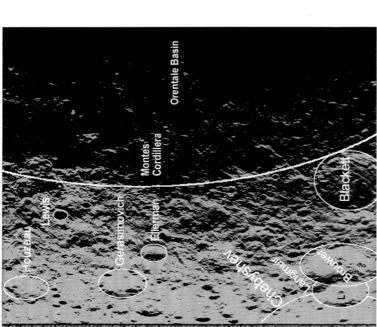

LO5-022H2

Sun Elevation: 5.34°

Distance: 5441.61 km

Key

Pre-Nectarian
Blackett, N?, 141 km
Houzeau, 71 km

Nectarian
Brouwer, EI?, 158 km
Chebyshev, EI?, 178 km

Early Imbrian
Montes Cordillera, 930 km
Orientale Basin, 930 km

This image shows the southwest ejecta blanket of the Orientale Basin. The heavy, continuous part of the ejecta blanket ends as it approaches Chebyshev, Langmuir, and Brouwer. The low sun elevation shows the stringy nature of its texture. Blackett has been covered with continuous ejecta from more than one source; there are also striations from the earlier Mendel-Rydberg Basin (next page). The apparent bending of the Orientale striations on the floor of Blackett is probably due to the oblique view of this deep crater.

South Pole-Aitken

SPA Rim / From Orientale

Buffon
Stetson
Leavitt
Mendel
Karrer
Riedel
Corl
Baldet
Stoney
Minkowski
Fizeau
Eijkman
Lippmann

South Pole-Aitken

LO5-022H1

Sun Elevation: 5°

Distance: ~5440 km

The Mendel-Rydberg Basin is to the southeast of Mendel. Some evidence of its ejecta is here, underneath the more obvious Orientale ejecta. The continuous Orientale ejecta blanket extends to Lippman, Fizeau, Karrer, and Buffon. The low albedo to the lower left, within the South Pole-Aitken Basin between Lippmann and Stoney has been confirmed by the Clementine mission. It is not so dark as mare in the floors of Karrer and Baldett and may have had its albedo raised by powdered ejecta from Orientale.

Key

Pre-Nectarian
Lippman, 180 km
Minkowski, 113 km

Nectarian
Buffon, 106 km
Eijkman, 54 km
Karrer, 51 km
Mendel, El?, 138 km
Mendel-Rydberg Basin, 630 km

Late Imbrian
Fizeau, 111 km
Stetson, 64 km

Chapter 11
The North Polar Far Side Region

11.1. Overview

This region extends from the North Pole of the Moon down to 55° north latitude (on the far side).

Lunar Orbiter spacecraft on missions 4 and 5 photographed this region with their medium-resolution cameras from the near side, shooting over the North Pole. As a result, the Lunar Orbiter coverage of the North Polar Far Side Region is quite low in resolution and at a very oblique angle.

The Clementine spacecraft, in its nearly polar orbit, systematically covered nearly all the Moon, including this region. While the topographic information in the Clementine photography of the equatorial area is poor (the mission was optimized for multispectral photography), the images north of 55° have a low sun elevation and are superior to the Lunar Orbiter photographs. Therefore, our comprehensive coverage of this region is primarily with the Clementine images. The Lunar Orbiter photographs of the region can be found on the enclosed CD, as well as the images from Clementine.

The elemental abundance within this area is fairly uniform except for a relatively high concentration of hydrogen near the North Pole, detected by the Prospector mission. The hydrogen may be at least partly in the form of water ice that has been deposited by comets. Moderately high iron content is associated with some of the deeper craters. The crater Compton on the southern boundary of this region near the western far side limb exhibits mare material on its floor. Compton also has a high level of thorium, as measured by Prospector.

The Humboldtianum Basin, just over the limb from this region, has spread the eastern part of its ejecta blanket onto part of this region. The Birkhoff and Coulomb-Sarton Basins have also spread their ejecta into the southeastern part of this region.

11.2. The Western Far Side Limb (Eastern Near Side Limb)

Early in its mission, Lunar Orbiter 4 took a photo of the northern sector of the East limb (eastern near side limb) of the Moon (see Figure 11.1). In the foreground is Mare Humboldtianum, within the inner ring of the Nectarian Humboldtianum Basin.

Schwartzschild is judged to precede the Humboldtianum Basin; therefore, its ejecta is covered by that of Humboldtianum. The Bel'kovich impactor landed just within the rim of Humboldtianum and the impact added its ejecta to the area. Below those layers is a fourth layer from the South Pole-Aitken Basin. These layers, deepest in the space between Bel'kovich and Shwarzschild, are analogous to multiple layers of sedimentary and volcanic deposits on Earth, but have been deposited by sudden, discrete impact events instead of by repeated episodes of flooding by water or lava from the same volcano.

11.3. Bel'kovich K, a Copernican Crater

In the part of the North Polar Far Side Region within 40° of the pole, only Bel'kovich K (Figure 11.2) and Birkhoffz have been identified as Copernican in age. In the area of the near side within 40° of the North Pole, Carpenter, Philolaus, Anaxagoras, Thales, and Hayn have been assigned to that period. A similar pattern occurs in the equivalent area near the South Pole, where no Copernican craters have been identified on the far side but Zucchius and Rutherford are near side Copernican craters.

This pattern suggests that the Moon reached its approximate current polar axis and rotational orientation before the Copernican Period. The reason for the relative lack of polar craters is that both comets and asteroids are likely to approach in the plane of the solar system, approximately aligned with the current lunar equator. Further, due to the focusing effect of Earth's gravity field, impacts are more likely to strike the near side of the Moon.

11.4. Clementine Mosaics

The approximate coverage of the Clementine images for the North Polar Far Side Region is shown in Figure 11.3. The area is covered in five sectors: the polar cap from the pole down to 70° north latitude and four sectors from 75° down to 55° north latitude, each covering a range of 45° in longitude. These sectors are based on Clementine, 750 nm data (NRL, USGS Map-a-Planet web page).

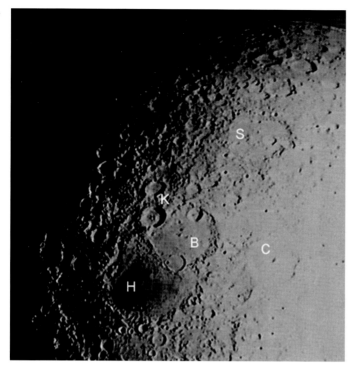

Figure 11.1. This Lunar Orbiter view is over the North Pole from the near side. It shows the Humboldtianum Basin (**H**) on the near side and Bel'kovich (**B**), Bel'kovich K (**K**), Schwarzschild (**S**), and Compton (**C**) on the far side. Part of LO4-023M.

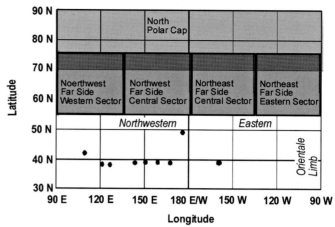

Figure 11.3. This figure labels the area of coverage of the Clementine images included in this chapter.

The images are shown starting on the next page. There are two pages for each image, one with the features identified and one clear. The first image is the polar cap and the next four images show the areas from 75° north down to 55° north latitude, starting from 90° east longitude and proceeding across the far side to 90° west longitude.

Figure 11.2. Copernican crater Bel'kovich K, 47 km in diameter, is centered at 63.8° north and 93.6° east. Clementine, NRL, USGS Map-a-Planet.

North Polar Cap
The North Pole to 70°N

Key

Pre-Nectarian Period
Brianchon, N², 134 km
Hermite, 104 km
Hippocrates, 60 km
Rozhdestvensky, 177 km
Seares, 110 km
Thiessen, 66 km

Nectarian Period
Froelich, 58 km
Nansen, PN², 104 km
Merrill, 57 km
Mezentsev, EI², 89 km
Milankovic, 101 km
Niepce, 57 km
Roberts, 89 km
Schwarzschild, 212 km
Shi Shen, 43 km

Early Imbrian Period
Heymans, 50 km
Karpinsky, 9 km

Late Imbrian Period
Lovelace, 54 km
Plaskett, 109 km

Eratosthenian Period
Ricco, 65 km

The far side is in the lower half of this image; the near side is on the upper half. The craters listed by period in the key above are all on the far side.

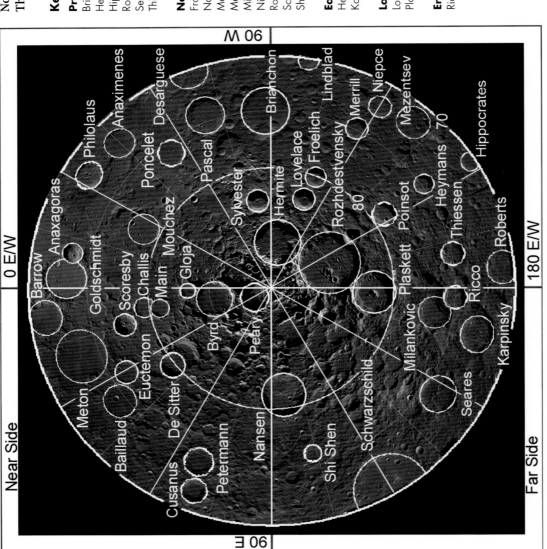

Near Side

Far Side

This north polar cap base image is an orthographic projection from the pole to 70° north latitude. It is derived from Clementine visual data from the NRL. This image, processed by USGS, is from the JPL web page.

North Polar Cap

The North Pole is within the cavity of the Near Side Megabasin and that of the South Pole-Aitken Basin. The Birkhoff ejecta blanket is in the eastern far side, below 80° north latitude. Ejecta from the Humboldtianum Basin softened the limb near 90° east longitude. The area from 135° to 180° east longitude is relatively free of basin ejecta later than that of the South Pole-Aitken Basin and therefore displays Nectarian craters Milankovic and Seares clearly.

The topography within 5° of the North Pole is dramatically lighted by the low sun angle, making it appear more rugged. It also is actually less smoothed by ejecta because the pole is near the edge of the ejecta blankets described above.

The crater walls that were shadowed during the Clementine mission are illuminated at other times of the year, because of the 1.5° tilt of the lunar axis relative to the plane of the Earth's orbit around the sun. There are no large areas of permanent shadow near the North Pole.

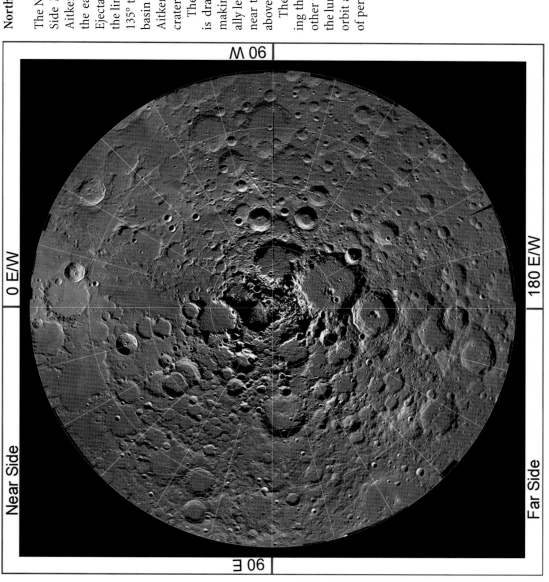

Near Side

0 E/W

90 W

180 E/W

90 E

Far Side

Northwest Far Side:
Western Sector

Key

Nectarian
Schwarzschild, 212 km
Volterra, PN?, 52 km

Early Imbrian
Compton, 162 km

Copernican
Bel'kovich K, 47 km

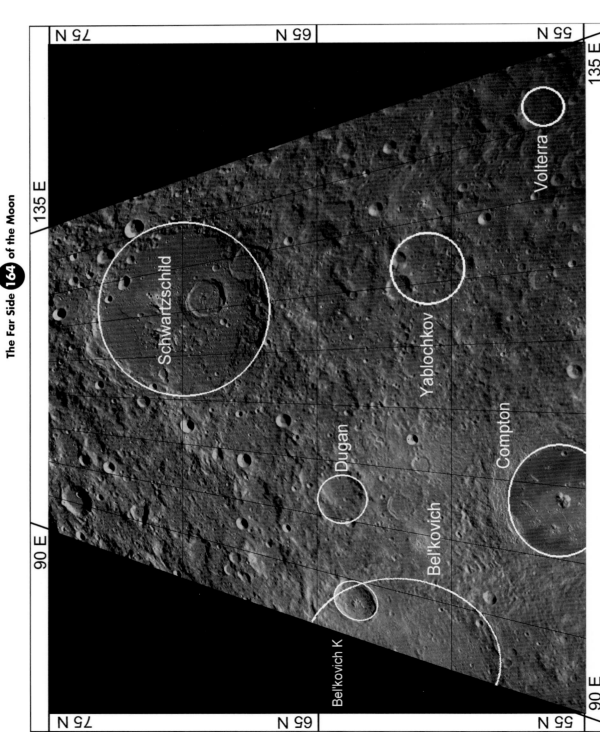

Schwartzschild arrived in the Nectarian Period, somewhat later than the formation of the Humboldtianum Basin. Schwartzschild rests on the rim and ejecta blanket of that basin and has obscured its rim. The sharp, young crater in the northern floor of Belkovich, designated as Belkovich K, is the only crater assigned to the Copernican Period on the far side within 30° of either the North or South Poles. Compton is a feature of the Early Imbrian Period. Note that few craters have impacted either its floor or extensive ejecta blanket.

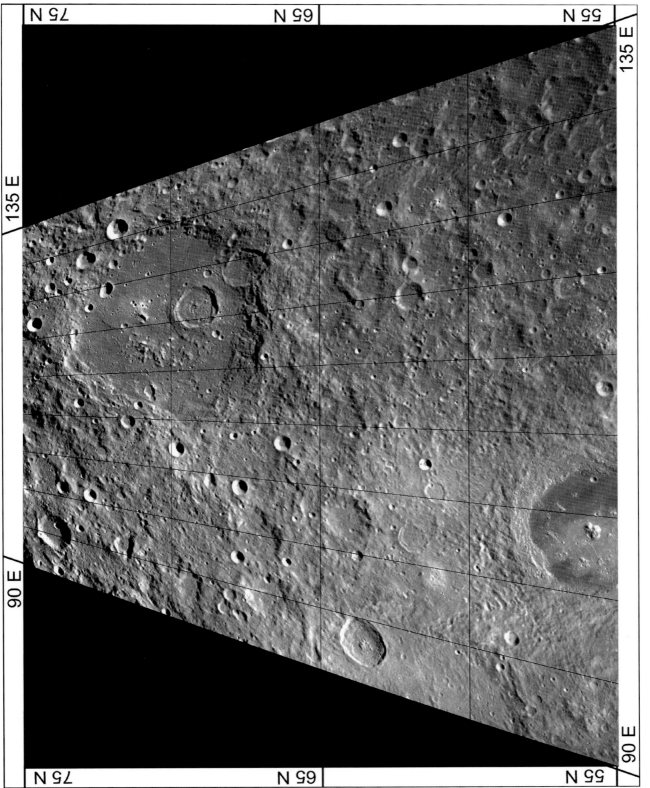

Northwest Far Side: Western Sector, 55° north to 75° north, 90° east to 135° east

The North Polar 165 Far Side Region

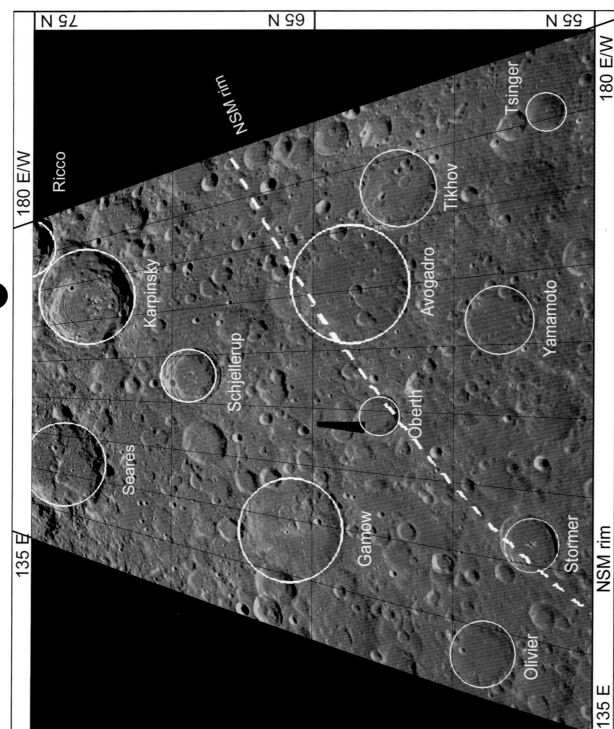

Northwest Far Side:
Central Sector

Pre-Nectarian Period
Avogadro, 139 km
Gamow, 129 km
Olivier, 69 km
Seares, 110 km
Stormer, 69 km
Tikhov, 83 km
Tsinger, 44 km
Yamamoto, 76 km

Nectarian Period
Oberth, 60 km

Early Imbrian Period
Karpinsky, 92 km

Late Imbrian Period
Schjellerup, 62 km

Eratosthenian Period
Ricco, 65 km

The depression of the Near Side Megabasin is north of the rim line shown here. The rim of the Near Side Megabasin bends away from the basin's center (on the near side) because the basin covers more than half of the Moon's globe. Relative to other sectors, this one is free of basin ejecta younger than that of the South Pole-Aitken Basin, and

The youngest crater here is Ricco, assigned to the Eratosthenian Period. Schjellerup is older, assigned to the Late Imbrian Period. Karpinsky is even older, of the Early Imbrian Period.

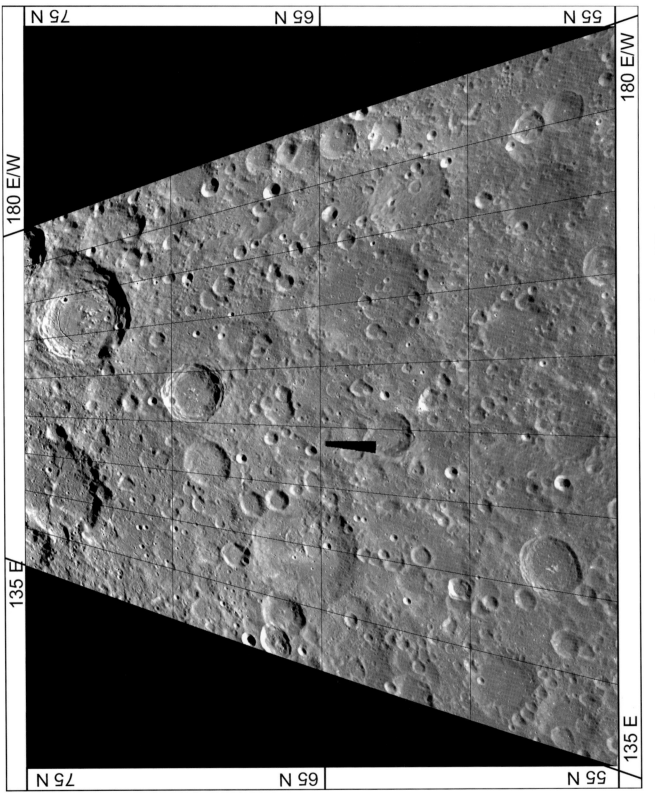

180 E/W

180 E/W

135 E

135 E

Northwest Far Side: Central Sector, 55° north to 75° north, 135° east to 180° east/west

The North Polar 167 Far Side Region

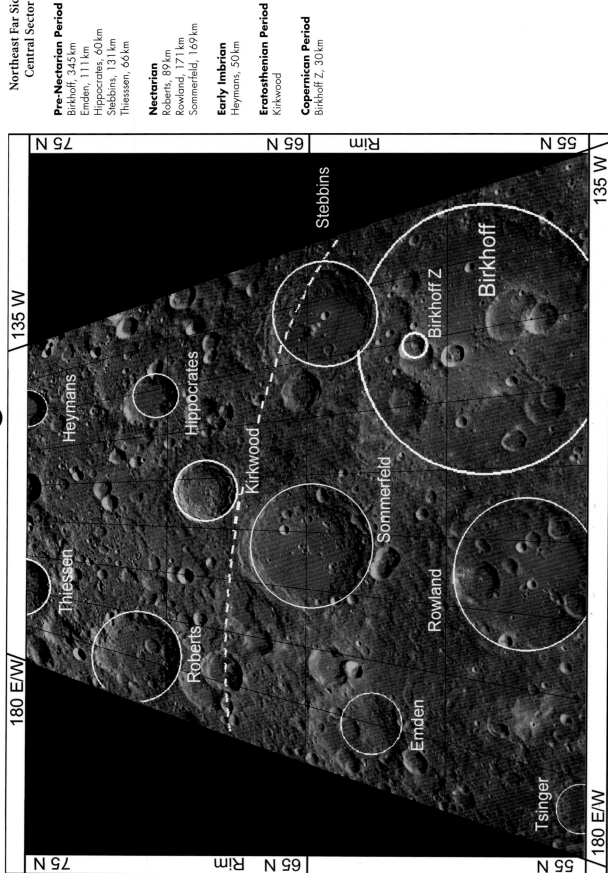

Northeast Far Side:
Central Sector

Pre-Nectarian Period
Birkhoff, 345 km
Emden, 111 km
Hippocrates, 60 km
Stebbins, 131 km
Thiesssen, 66 km

Nectarian
Roberts, 89 km
Rowland, 171 km
Sommerfeld, 169 km

Early Imbrian
Heymans, 50 km

Eratosthenian Period
Kirkwood

Copernican Period
Birkhoff Z, 30 km

The northern edge of the Near Side Megabasin continues here. Because of scalloping, the actual rim may be as much as 5° of latitude away from the dotted line, the elliptic edge predicted by the model based on the Clementine elevation data.

Stebbins is clearly younger than Birkhoff, although both are of the pre-Nectarian Period. Birkhoff Z, the sharp crater on the floor of Birkhoff near Stebbins, is assigned to the Copernican Period. An even younger small crater on its floor causes the brightness that could be mistaken for a peak.

Northeast Far Side: Central Sector; 55°north to 75°north, 135°west to 180°east/west

The North Polar 169 **Far Side Region**

Northeast Far Side:
Eastern Sector

Pre-Nectarian Period
Brianchon, 134 km
Omar Khayyam, 70 km
Poczobutt, 195 km
Van't Hoff, 92 km

Nectarian Period
Coulomb, 89 km
Cremona, 85 km
Dyson, 63 km
Merrill, 57 km
Mezentsev, El?, 89 km
Niepce, 57 km
Paneth, 85 km
Smoluchowski, 83 km
Zsigmondy, 65 km

Early Imbrian Period
Ellison, 36 km
Noether, 67 km

Late Imbrian Period
Lindblad F, 42 km

The ejecta blanket of the Imbrium Basin has covered nearly all of this sector, except for a few degrees on the western (left) edge and the lower left corner, roughly the area beyond the rim of the Near Side Megabasin (by concidence). Imbrium ejecta softened the shapes of many of the craters that are older than Imbrium, obscuring evidence for exact ages. Striations from Van't Hoff, a Nectarian crater, can be seen exposed where the Imbrium ejecta blanket has not covered them. Further, Lindbad F has impacted the Imbrium ejecta blanket after it was deposited.

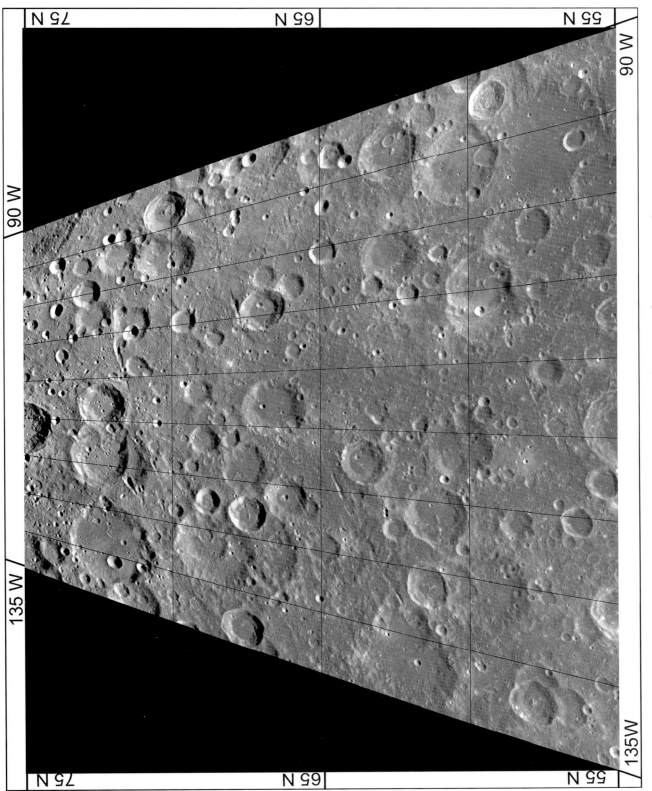

Northeast Far Side: Eastern Sector; 55° north to 75° north, 90° west to 135° west

The North Polar **171** **Far Side Region**

Chapter 12

The Orientale Limb Region

12.1. Overview

The last region to be covered in this book is the West Limb, the eastern boundary of the far side dominated by the Orientale Basin (Figure 12.1). This completes the photographic coverage of the entire Moon that was begun in the companion volume, "Lunar Orbiter Photographic Atlas of the Near Side of the Moon," which started with the near side portion of the same area. Orientale, the youngest of the large basins, the one that ended the era of the basins, is on the leading edge of the Moon as it rotates around the Earth. This large basin and its ejecta establish the boundary in time between the Early Imbrian Period and the Late Imbrian Period.

The eastern edge of the Orientale Basin can be seen from Earth, but only when the Moon's libration is favorable. Its eastern ejecta pattern was detected first, and called the Hevelius formation after the seventeenth century astronomer. Hevelius started the practice of using names with roots meaning "eastern" for features near the western edge of the near side because he was observing through a telescope whose optics inverted images.

At about 25° north latitude along the limb, the Orientale ejecta blanket terminates, exposing the ejecta of the near side Imbrium Basin. The Lorentz Basin (pre-Nectarian, 325 km) lies beneath that. The shore of Oceanus Procellarum comes to within 10° of the far side in the range of 20°–50° north latitude.

Near the limb south of Orientale are the Mendel-Rydberg and Pingre-Hausen Basins. First came Mendel-Rydberg (630 km) in the pre-Nectarian Period, followed by Pingre-Hausen (300 km) in the Nectarian Period. These two overlapping basins were then covered with Orientale ejecta.

12.2. The Crater Hausen

Hausen (Figure 12.2, Eratosthenian, 167 km) is the third largest impact feature to have ever followed Oriental, the last of the basins. Of craters that were formed after the Orientale event, only Humboldt (189 km) and Tsiolkovskiy (185 km), both Late Imbrian, are larger than Hausen.

The sudden shift in the size of impact features between the 950 km Orientale and the 189 km Humboldt suggests an abrupt end of the time of large impactors.

12.3. Radial Profile of the Orientale Basin

The general shape of the Orientale Basin can be estimated from the photographs. Photoclinometry, the science of inferring slopes from brightness variations, can provide profiles in the east–west direction. But these techniques emphasize irregularity in the basically symmetric result of the impact. A radial profile derived by averaging the elevations at each radius from the center is shown in Figure 12.3. This type of profile reduces the random variations introduced by the inherently chaotic impact process. The resulting graph is typical of the profiles of all impact basins, although there are variations in the number of internal rings (2, 1, or 0) and the degree of mare flooding.

12.4. Photos

Lunar Orbiter Medium-Resolution Frames

The Orientale Limb Region is covered in three frames and two mosaics of medium-resolution frames from Lunar Orbiter 4, presented from north to south. The principal ground points of the medium-resolution photos of this region are shown in Figure 12.4. Frames LO4-181M and -187M have been combined in a mosaic, as have the three photos LO4-180M, LO4-186M, and LO4-194M.

The medium-resolution photos are shown starting on page 174. All of the medium-resolution photos are shown before the high-resolution frames, to avoid breaking the continuity of contiguous coverage of the medium-resolution photos.

Lunar Orbiter High-Resolution Subframes

There are three high-resolution subframes for each of the medium-resolution principal ground points that are labeled in Figure 12.4. In addition, there are four subframes, LO4-196H3, LO4-195H1, LO4-195H2, and LO4-195H3, whose medium-resolution frames are unusable. In general, subframes overlap in the east–west direction, and are presented as mosaics. However, the subframes of LO4-189H, H2, and H3 and LO4-188 H1 and H2 are shown on separate pages.

Figure 12.1. The 950 km Orientale Basin is centered on the far side (20° south, 95° west). This young basin is the archetype for all ringed basins of the Moon. Part of LO4-187M and LO4-181M.

Figure 12.2. The 167 km Hausen crater (65° south, 88.1° west) is shown at the same scale as the Orientale Basin (Figure 12.1). It is only 15% smaller than Humboldt, the largest crater formed after the Orientale event. Part of LO4-186M.

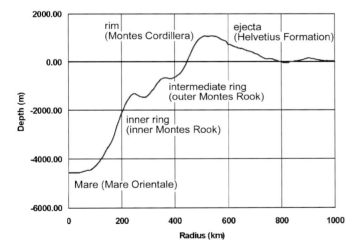

Figure 12.3. Radial profile of the Orientale Basin, showing the mare, inner ring, intermediate ring, the main ring (rim), and the ejecta blanket. The slope and estimated curvature of the preimpact surface have been removed in this graph. This profile was derived from the Clementine elevation map (Topogrd2, 0.25° resolution; Zuber, 2004). The elevations were averaged over azimuth as a function of radius.

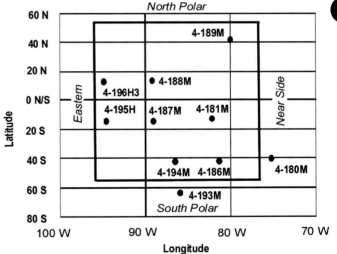

Figure 12.4. This figure labels the principal ground points of the medium-resolution Lunar Orbiter photos included in this chapter. The principal ground points are also shown for LO4-196H3 and LO4-195H, which are combined in a high-resolution mosaic.

For the Lunar Orbiter 4 photos of this chapter, the high-resolution subframe H2 is centered on the principal ground point, H3 is to the north of H2, and H1 is to the south of H2.

When the notes accompanying the high-resolution subframes mention specific features, they specify code letters that appear in the images.

LO4-189M   Sun Elevation: 18.36°   Distance: 2882.19km

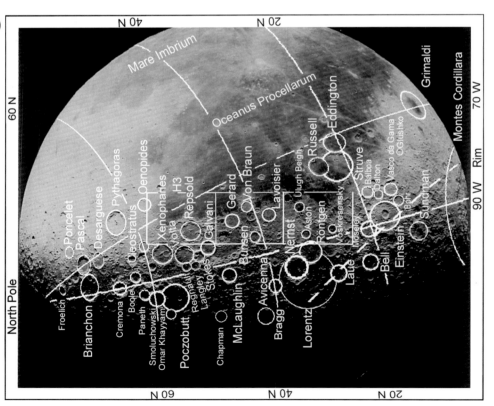

The striated ejecta blanket of the Orientale Basin extends from Montes Cordillara to about Röntgen, near 30° north latitude. North of there, the ejecta blanket of Imbrium is exposed where it has not been flooded with mare.

Key for far side features above 40°:

Pre-Nectectarian
Avicenna, 74 km
Bragg, 84 km
Chapman, 71 km
McLaughlin, №, 79 km
Onar Khayyam, 70 km

Nectarian
Paneth, 65 km
Poczobutt, 195 km
Röntgen, 126 km

Cremona, 85 km
Smoluchowski, 21 km

Distance: 2677.28 km

Sun Elevation: 14.52°

LO4-188M

Orientale's ejecta blanket has erased or grossly modified nearly all sizable craters in this area, evidence for the decline in the rate of large impact events after the Orientale event.

Key to far side features from 0° north/south to 40° north

Pre-Nectarian
Helberg, 62 km
Laue, 87 km
Lorentz Basin, 312 km
Röntgen, 126 km
Sundman, N?, 40 km

Nectarian
Bell, El?, 86 km
Elvey, El?, 74 km
Nernst, 116 km
Nobel, El?, 48 km

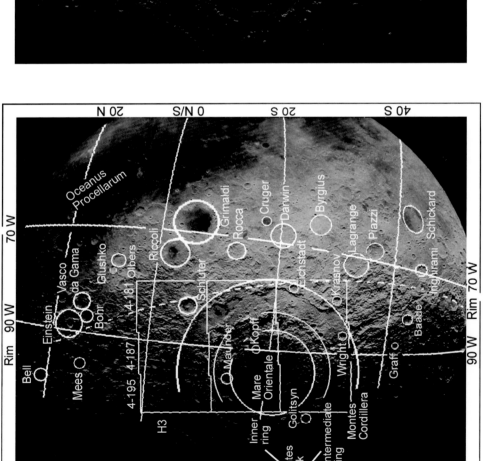

LO4-187M (left) and LO4-181M (right)

Sun Elevation: ~14.5°   Distance: ~2724km

Key to far side features from 40° south to 0° north/south

Pre-Nectarian
Kopf, Nº, 41 km

Early Imbrian
Orientale Basin, 930 km

Eratosthenian
Maunder, 55 km

Late Imbrian
Golitsyn, 36 km

This image shows the complex pattern of the Orientale Basin, the internal and external concentric rings and the striations that start at the main ring (Montes Cordillera) and extend to the edge of the ejecta blanket. The rim of the Near Side Megabasin (dashed line) passes through the region dominated by the Orientale Basin, raising the western part of that basin above its eastern part. Because the Near Side Megabasin covers more than half of the Moon, its rim curves away from its center.

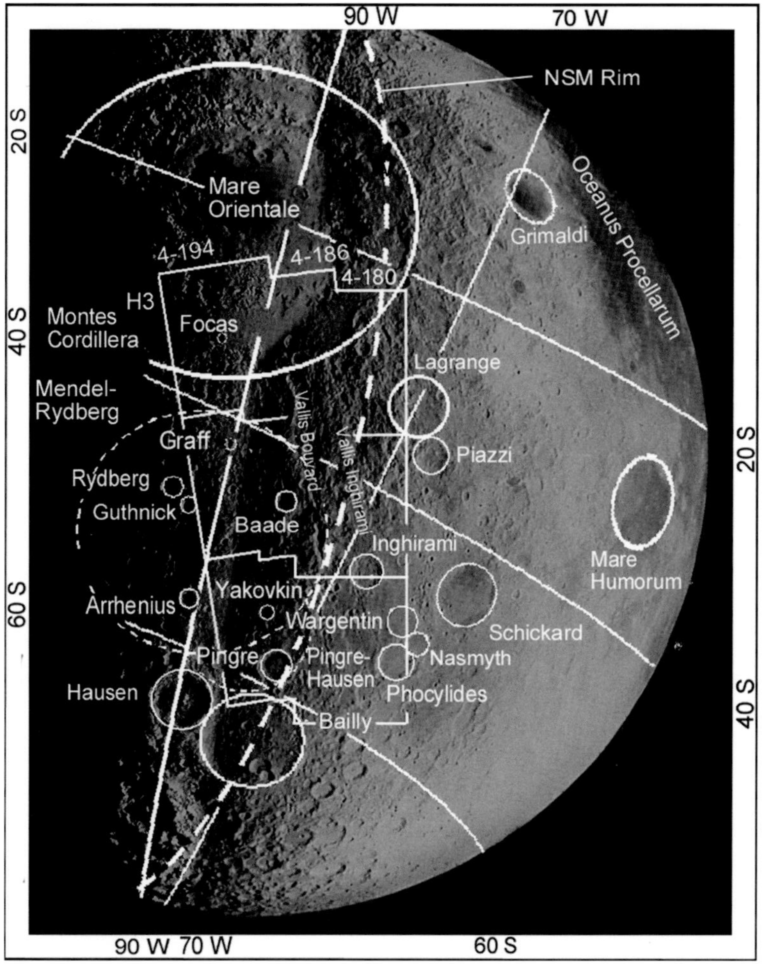
90 W
70 W
NSM Rim
20 S
Mare
Orientale
Grimaldi
Oceanus Procellarum
4-194
4-186
4-180
H3
Montes
Cordillera
Focas
40 S
Lagrange
Mendel-
Rydberg
Graff
Vallis Bouvard
Vallis Inghirami
Piazzi
20 S
Rydberg
Guthnick
Baade
Inghirami
Mare
Humorum
60 S
Yakovkin
Arrhenius
Wargentin
Schickard
Pingre
Pingre-
Hausen
Nasmyth
Hausen
Phocylides
40 S
Bailly
90 W 70 W
60 S

LO4-194M (left), LO4-186M (upper left), and LO4-180M

Sun Elevation: ~15°

Distance: ~3036 km

The apparent deep valley crossing this image is simply an artifact of the mosaic. The Mendel-Rydberg Basin is more than half as big as the Orientale Basin, but is much more degraded. The best photographic view of the rim of the Near Side Megabasin anywhere is in this image, between Inghirami and the western Montes Cordellaris.

Key to far side features below 40° south

Pre-Nectarian
Pingre-Hausen Basin, 300 km

Nectarian
Mendel-Rydberg Basin, 630 km

Late Imbrian
Graff, 36 km

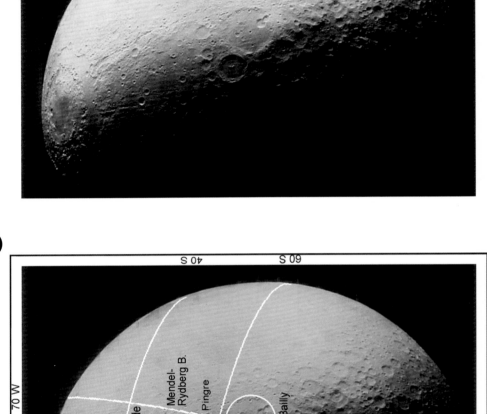

LO4-193M

Sun Elevation: 10.31°

Distance: 3557.30km

The area between the Orientale Basin and Bailly is rich in impact features younger than the Orientale Basin; their ejecta patterns lie on top of that from Orientale, and could be easily sampled. The rim of the proposed Near Side Megabasin would provide relatively pristine crust as the target material for these features.

Key

Nectarian
Bailly, 287 km
Pingre, 88 km

Early Imbrian
Fenyi, 38 km
Orientale Basin, 930 km

Late Imbrian
Baade, 55 km
Chant, 33 km
Steklov, 36 km

Erastosthenian
Hausen, 167 km
Rydberg, 49 km

Copernican
Guthnik, 36 km

LO4-189H3   Sun Elevation: 18.36°   Distance: 2882.19 km

The features in this image are all on the near side, but it introduces the boundary between Oceanus Procellarum and the rising crust within the Near Side Megabasin. The interplay between the mare flood and the earlier cratered features is clearly illustrated by the stress cracks in Repsold (**R**, 109 km). These cracks were probably opened by contraction of the basaltic mare as it cooled. Further spreading of fractures may be due to upward pressure by intrusion of magma beneath Repsold.

LO4-189H2   Sun Elevation: 18.36°   Distance: 2882.19 km

The area around and to the northwest of Gerard (**G**, 90 km) is somewhat puzzling. Gerard itself may be two or three separate craters that have left an overlapping pattern. Further, there may be a very old pre-Nectarian basin whose center is just beyond the upper left corner of this subframe. The entire area appears to have been flooded with mare, but at an older time than the fresh mare surface of Oceanus Procellarum to the east. Von Braun (**VB**, 22 km) is named after the engineer who led the Saturn 5 project, which brought men to the Moon.

LO4-189H1   Sun Elevation: 18.36°   Distance: 2882.19 km

The dark area between Aston (**A**, 43 km) and Voskresen-sky (**V**, 49 km) is mapped as Imbrium ejecta, but seems to have the characteristics of a dark plains deposit, sometimes seen on the margins of maria on the near side. Voskresensky shows some flooding by mare mate-rial, which may be just beneath the surface here. There are signs that the upward pressure of lava has lifted a melt sheet on the floor of Voskresensky. The ray passing south of Voskresensky has come from Ohm, far to the southwest.

LO4-196H3 (left) and LO4-188H3   Sun Elevation: ~15°   Distance: ~2677 km

This is a mosaic of adjacent subframes, spanning from 98° west longitude to 85° west longitude. The uniform sizes of the small craters indicate they are secondaries from the Orientale Basin. The field of secondaries surrounding Orientale is called the Outer Hevelius Formation. The secondary impactors must have been coherent bodies whose diameters were of the order of 1 km in order to form these craters, whose typical diameter is about 10 km. The scale of this image can be determined from the size of Bell (**B**, 86 km). Bell has received some of these Orientale secondaries, but also has older craters that show more erosion. It is assigned to the Nectarian Period, but could be either pre-Nectarian or Early Imbrian.

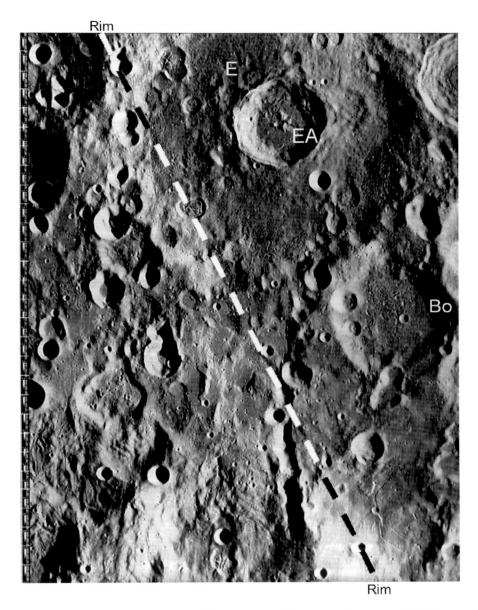

Rim

E

EA

Bo

Rim

LO4-188H2   Sun Elevation: 14.52°   Distance: 2677.28 km

The large crater in the top of this subframe has been named to honor Albert Einstein (**E**, 198 km). Bohr (**Bo**, 71 km) is younger, of the Nectarian Period. Einstein A (**EA**, 51 km) appears free of Orientale ejecta, but a few substantial craters have impacted it, indicating it was formed in the Late Imbrian Period. The edge of the Near Side Megabasin runs through this image (dashed line). To the right is the interior of that basin and to the left is the fragmented, porous rim. Although there are no identifiable features of the rim here, the upward slope of the rim causes the greater average brightness west of the dashed line.

LO4-188H1   Sun Elevation: 14.52°   Distance: 2677.28 km

This entire subframe shows a sector of the ejecta blanket of the Orientale Basin, called the Inner Hevelius Formation. The northern arc of Montes Cordillera runs just below the bottom edge of this image. The chaotic radial pattern of the ridges and troughs is clearly illuminated by the low sun angle. Faint outlines of craters that have been buried by the ejecta have survived. Maunder A (**MA**, 15 km) is the only feature available for scale here.

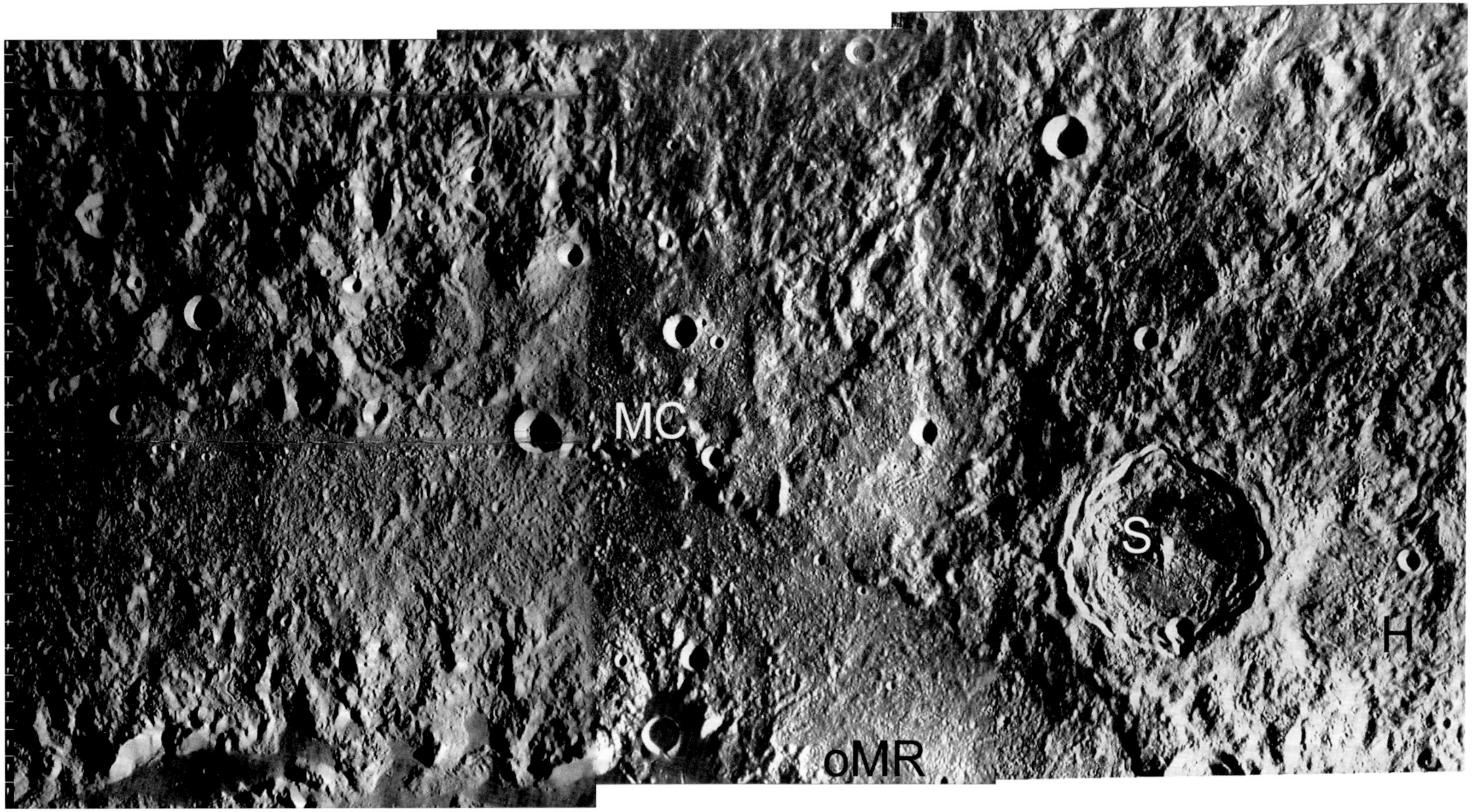

LO4-195H3 (left), LO4-187H3 (center), and LO4-181H3 (right)

Sun Elevation: ~14.5°

Distance: ~2724 km

A circumferential plain, probably a melt sheet, fills the sloped basin floor between the outer Montes Rook (**oMR**, 620 km) and the north-eastern sector of Montes Cordillera (**MC**, 930 km). Hartwig (**H**, 79 km), just to the east of Schluter (**S**, 89 km) and just a little smaller, has been covered with Orientale ejecta. Schluter's age is assigned to the Late Imbrian Period because it is free of ejecta from the Orientale event, but was there before the period of mare emplacement, which has extruded into its northern floor.

The Orientale **185** Limb Region

LO4-195H2 (left), LO4-187H2 (center), and LO4-181H2 (right)

Sun Elevation: ~14.5°

Distance: ~2724 km

Mare Orientale (**MO**) covers the central floor of the Orientale Basin, surrounded by the inner Montes Rook (**iMR**, 480 km), the outer Montes Rook (**oMR**, 620 km), the circumferential melt sheet beyond the outer Montes Rook, and Montes Cordillera (**MC**, 930 km) beyond the melt sheet. The circumferential melt sheet has received its own intrusion of mare lava in places such as Lacus Autumnis (**LA**). The shores of mare flows such as Lacus Autumnis, Lacus Veris (**LV**), and Mare Orientale rise around their boundaries, a sign that the mare has risen above its current level and then subsided as it cooled.

LO4-195H1 (left), LO4-187H1 (center), and LO4-181H1 (right)

Sun Elevation: ~14.5°   Distance: ~2724 km

In the southeastern sector of the interior of the Orientale Basin, the melt sheet between the outer Montes Rook (oMR) and Montes Cordillera (MC) widens. Patterns of stress cracks (sc) appear on the melt sheet on the slope leading from the inner Montes Rook down to Mare Orientale. A fine radial pattern appears on the circumferential melt ring in this sector, not particularly correlated with the deeper pattern outside Montes Cordillera. Rimae Pettit (RP, 450 km), radial narrow depressions, extend from the mare across the Montes Rook and across the melt sheet. These features may be a result of the formation of Montes Rook, which may have occurred shortly after the primary impact event, as a result of rebound of the compressed material in the transient crater. Rimae Focas (RF, 100 km), narrow depressions between the inner and outer Montes Rook, are filled with lava. Pettit (P, 35 km) and Nicholson (N, 38 km) may have come from a pair of impactors. Fryxell (F, 18 km) and Shuleykin (Sh, 15 km) have landed on the inner Montes Rook.

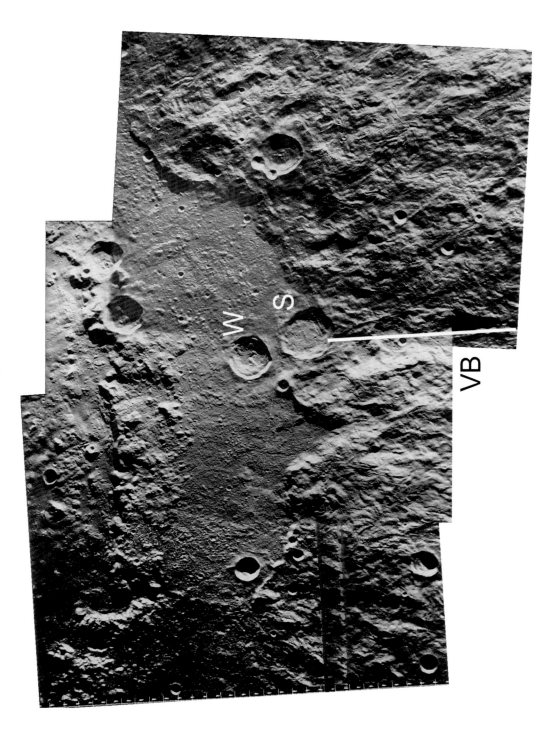

LO4-194H3 (left), LO4-186H3 (center), and LO4-180H3 (right)

Sun Elevation: ~15°

Distance: ~3036 km

This shows the southern sector of the outer Montes Rook, the wide circumferential melt sheet, and Montes Cordillera. The start of the deep trough called Vallis Bouvard (VB) extends south from Wright (W, 39 km), the young crater of the Eratosthenian Period and the more eroded crater Shaler (S, 48 km), of the Late Imbrian Period.

LO4-194H2 (left), LO4-186H2 (center), and LO4-180H2 (right)   Sun Elevation: ~15°   Distance: ~3036 km

The prominent trough and ridge system of Vallis Bouvard (VB) continues southward, crossing this entire image. The length of Vallis Bouvard is estimated to be 284 km. Baade (B, 55 km) has impacted the ridge on the western boundary of Vallis Bouvard. Vallis Inghirami (VI) is to the east of Vallis Bouvard and extends for 148 km.

The Orientale 189 **Limb Region**

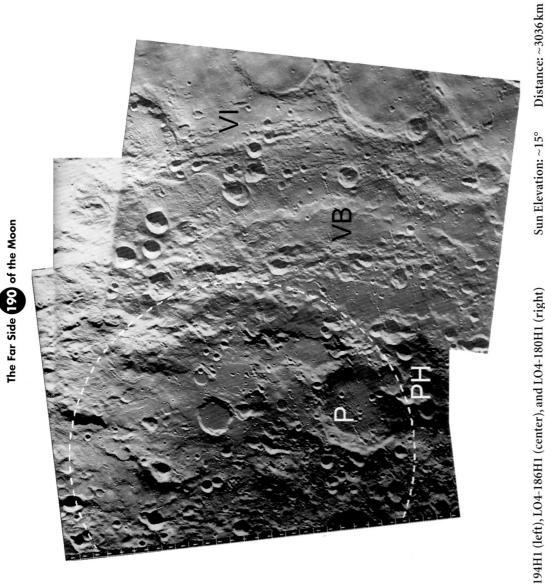

LO4-194H1 (left), LO4-186H1 (center), and LO4-180H1 (right)   Sun Elevation: ~15°   Distance: ~3036km

Vallis Bouvard (**VB**) and Vallis Inghirami (**VI**) change their shape here; the continuous valleys of the northern images break into irregular patterns of secondary craters, no longer really valleys. The radial pattern of the continuous ejecta blanket of Orientale thins out in this image, terminating near Pingre (**P**, 88 km). Pingre itself has escaped the heavy blanket, but has received a powdery dusting of finely divided plains forming material from Orientale. Pingre has landed on the floor of the Pingre-Hausen Basin (**PH**, 300 km), formed in the pre-Nectarian Period.

LO4-193H3   Sun Elevation: 10.31°   Distance: 3557.30 km

The smooth patch (**sp**) appears to have been deposited as molten material. It may be an exposed part of the floor of the Mendel-Rydberg Basin or may be Orientale ejecta. This may be a suitable landing site for examination of the minerals in this area, but higher resolution photography would be required. Baade (**B**, 55 km) is an example of an elliptical crater, an indication that its impactor arrived at a low elevation. The alignment of its major axis with Orientale ejecta suggests that it is a secondary crater, formed by a block ejected from Orientale.

LO4-193H2   Sun Elevation: 10.31°   Distance: 3557.30 km

The crater Hausen (**H**, 167 km) is of the Erastosthenian period. It has deposited some of its ejecta on the floor of its Nectarian neighbor Bailly (**B**, 287 km). This juxtaposition provides an opportunity to study the differences in texture between impact features of these two periods. Note the relative frequency of craters on the floors, rims, and ejecta blankets of the two features. Also note the collapse of terraces in the wall of Bailly's rim.

LO4-193H1   Sun Elevation: 10.31°   Distance: 3557.30 km

Drygalsky (**D**, 149 km) can be seen to be pre-Nectarian, as indicated by the many craters of considerable size that have impacted it. Yet it retains its basic shape, includ- ing the large central peak. Not enough detail is visible in Boltzmann (**B**, 76 km) to establish its age.

Chapter 13

The Near Side Megabasin

13.1. How the Far Side Differs from the Near Side and Why

Chapter 1 presented the history of the discovery of major differences between the near side and the far side of the Moon, based on images returned from spacecraft. In this chapter, additional evidence of difference is provided and a new view of the cause of the difference is discussed. There is good evidence that very early in the history of the Moon, a giant impact created the Near Side Megabasin. The near side is nearly all within the depression of this giant basin, and its rim and ejecta cover the far side.

A photograph of the eastern limb from the Apollo camera in the Command Module of Apollo mission 16, taken as it awaited return of the Lunar Module ascent stage, is shown in Figure 13.1. The terminator is about 35 beyond the limb in this picture. Although some of the impression of ruggedness on the far side is due to its nearness to the terminator (where the low sun angle increases the length of shadows), most of the impression is due to differences in the topography.

The Clementine and Lunar Prospector spacecraft missions gave us a wealth of new information about the Moon, but the mystery deepened. Clementine's laser altimeter revealed the remarkable depth of the giant basin in the southern far side, the South Pole-Aitken Basin, which extends from the South Pole up to 17° south latitude. It is about 6.8 km deep, deeper than any other lunar crater but mare floods only the deepest craters and basins within it. The altimeter also confirmed, with additional precision, that the Moon is irregularly shaped, with a bulge of several kilometers on its far side (see Figures 13.2 and 13.3). Further, its center of gravity (the center of spacecraft orbits) is shifted about 2 km toward the near side from the center of its surface figure.

The Lunar Prospector spacecraft provided additional news of the unusual surface distributions of relatively heavy elements like titanium, iron, and thorium. These materials tended to be more abundant at the surface of the near side. The only unusual concentrations on the far side were found to be within the South Pole-Aitken Basin.

We have known since 1984 that the Moon owes its origin to the Mars-sized planet Theia. The name refers to the Titan goddess in Greek mythology who created light and was the mother of Helios (the sun), Eo (dawn), and Selene, who personified the Moon. The orbit of Theia was disrupted until it struck Earth a glancing blow and was essentially vaporized, along with part of the Earth's crust. The debris formed a temporary ring around Earth. The Moon coalesced from this ring, having lost its volatile material to space and part of its iron core to Earth. So why did not the Moon form a nearly spherical, uniform ball, perhaps slightly ellipsoidal because of its spin about its poles?

We know why the Moon has a near and a far side; that is, why we consistently see the same side throughout the lunar month. Just as the Moon's gravity causes tides in the oceans of Earth, the Earth's much stronger gravity causes a tidal effect on the solid (once partially molten) Moon. While the Moon was still spinning in relation to Earth's gravity field, the resultant dissipation of energy damped the spin until the Moon fell into synchronism. That is, its period of rotation is equal to its period of revolution. This has happened with all the large moons of the planets. But until recently, there has been no explanation of the strongly asymmetric shape of the Moon.

Cadogan and Whitaker each proposed that a large impact had hit the near side, creating a basin much larger than the known basins and spreading its ejecta on the far side. Cadogan proposed the Gargantuan Basin and Whitaker proposed the Procellarum Basin. Both basins share the arc of the west edge of Oceanus Procellarum, but the proposed Procellarum Basin extended further to the east. Their thinking was essentially correct, but the geologic evidence of their proposed basin boundaries was not confirmed.

13.2. Impact Basin Models: The Search for the Source of the Far Side Crust

Recently, using the refined data from the Clementine laser altimeter that is shown in Figures 13.2 and 13.3, I undertook a search for a basin whose ejecta covers the far side of the Moon.

The first step was to form a generalized model of an impact basin. Projectiles from beyond Earth's orbit cause impact: maverick stony or metallic bodies from the asteroid belt or icy comets from the Kuiper belt. They arrive at the Moon with velocities far in excess of the speed of sound in

Figure 13.1. The eastern limb of the Moon, taken by the mapping camera in the Apollo 16 Command Module. Mare Smythii and, above that, Mare Marginis are mostly on the near side. The limb (90° east longitude) runs through the right hand edge of Mare Smythii, which is on the lunar equator. The far side is to the right. Photo a16_m_3021, NASA, LPI.

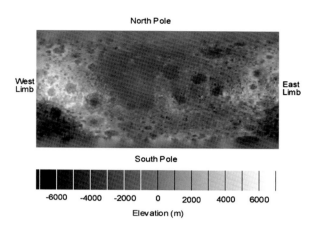

Figure 13.2. Elevation map of the Moon, centered on its near side. This geographic map displays elevation as a grayscale against latitude from 90° south to 90° north, and longitude from 180° west to 180° east. The data, called topogrd1, was derived from the Clementine LIDAR instrument. NRL, Zuber, 2004.

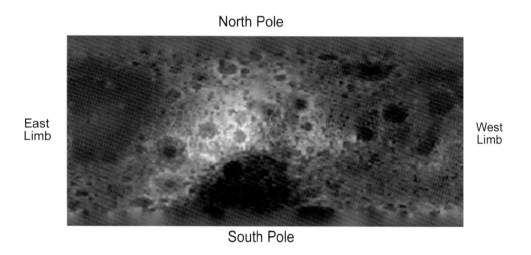

Figure 13.3. This elevation map is the same as that of Figure 13.2 except that it is centered on the far side. Note that the East Limb is the western limb of the far side. The dark depression is the South Pole-Aitken Basin. The large Korolev Basin is just north of the South Pole-Aitken Basin, at the top of the far side bulge. The bulge would be approximately circular if it were not for the impact of the South Pole-Aitken Basin.

the surface material, in other words at hypervelocity. In such circumstances, the effect is nearly the same as a powerful explosion. The most important property of the impactor is simply its kinetic energy. The angle of the impact is of secondary importance. Only if the approach is less than 30 from the horizontal does the shape of the impact basin become significantly elliptical, rather than circular.

Through a discipline called dimensional analysis, impacts of all sizes can be compared to normalized models. The technique covers size ranges from laboratory experiments all the way up to the large lunar basins. Figure 13.4 shows radial profiles of several lunar basins that are first measured using Clementine elevation data and then normalized by both

depth and diameter. It also shows a model radial profile that matches key features of the selected basins.

The depth and the diameter of the profiles were each normalized separately because the ratio of depth to diameter depends on many factors including the energy of the impact and the nature of the target surface. For example, the density, porosity, layering, and curvature of the target surface all influence how the energy of impact is dissipated. Further, isostatic compensation can act to decrease the vertical scale of a basin over time. This effect would be especially strong for a very large and very early basin, which is exactly what is being sought.

The model of an impact basin and its ejecta shown in Figure 13.4 fits the medium-sized basins, but could not be

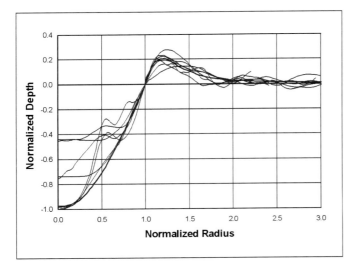

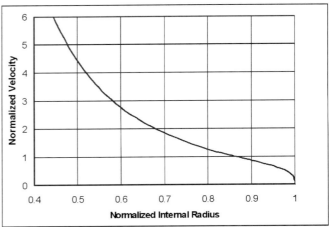

Figure 13.4. The *solid line* is a radial profile of a generalized impact basin. The *light lines* are radial profiles of several lunar basins ranging from 200 to 900 km in diameter. A radial profile of a basin is a plot of the average height or depth around a circle, as a function of the radius from the center of the basin. Some of these basins have inner and outer rings and have been partially filled with mare, features that are neglected in the model. The sample basins were measured using the high-resolution Clementine database topogrd2.

Figure 13.5. This radial profile of ejection velocity, together with the assumptions supported by dimensional analysis and experiment, produces the model ejecta pattern of Figure 13.4.

Now that I had a general model of an impact basin, I was ready to search for the basin that had left its debris on the far side.

expected to fit a basin of the size of the South Pole-Aitken Basin or larger because of the curvature of the Moon. For the smaller basins, the ejecta can be assumed to follow approximately ballistic trajectories, like the flight of a cannon ball. But actually, the ejected material is launched in an elliptical orbit into the vacuum of space. Therefore its launch velocity (speed and angle) must be estimated and included in the model.

How the ejected material is launched at the edge of the expanding explosion is known from laboratory measurements. The angle of launch is nearly constant at about 45 and the mass rate of ejection per angle of azimuth is nearly independent of radius. However, the velocity varies strongly with the radius of launch, being very high at the point of impact and falling to zero at what will become the edge of the impact basin. Figure 13.5 shows the radial profile of velocity that is consistent with the ejecta model of Figure 13.4. Recently, a theoretical study (Richardson, 2007) of the ejection velocity has been published that agrees well with the empirical curve of Figure 13.5, and agrees precisely out to a radius of 0.91.

Another important difference between the ejecta pattern of very large basins and medium basins is the area available for the circular ring of ejecta to occupy. On an approximately flat surface, the depth of a ring of ejecta at a given radius from the center is simply its volume divided by the width of the ring multiplied by the circumference of a circle of equivalent radius. But when the basin is so large that its ejecta must be considered as falling on a spherical surface, the circumference of the circle is reduced. In fact, when the ejecta, in its orbit, reaches the far side of the Moon (the antipode of the impact point), the ejecta from all directions converges, theoretically at a point. Of course, variations from the assumptions actually spread the ejecta out somewhat in the vicinity of the antipode.

13.3. The Near Side Megabasin and its Effects on the Far Side

I varied the model's parameters (latitude and longitude of the center, the diameter, and the depth) until I arrived at a best match to the measured lunar surface. The result was surprising. I found that a single new impact basin, together with the familiar South Pole-Aitken Basin, could account for all the mysterious evidences of topographic asymmetry. At this point, I did not consider the effect of isotopic adjustment. However, only the depth parameter of the basin is affected by that phenomenon.

Internal Basin and Rim
To do its work, the new impact basin had to be very large indeed. In fact, it covered more than half the Moon. Its center is on the near side at latitude 8.5° north and 22° east, in the western part of Mare Tranquillitatis, not far from the Apollo 11 landing. But its rim is on the far side! The radius of the Near Side Megabasin turned out to be about 3,000 km, more than half way around the Moon. The basin is somewhat elliptical, with an eccentricity of about 0.4. Its major axis runs from southeast to northwest.

All of the mare areas of the near side are within this basin. Since it includes many smaller basins, I refer to it as the Near Side Megabasin. The South Pole-Aitken Basin also has smaller basins within it and could be called the Far Side Megabasin if it did not already have a familiar name.

Ejecta Blanket
When the parameters of the Near Side Megabasin and the South Pole-Aitken Basin were adjusted, these two basins and their ejecta explained most of the evidence of the mysterious asymmetry of the Moon. The bulge on the far side is centered

on the antipode of the Near Side Megabasin. The South Pole-Aitken impact later took a big bite out of the ejecta of the Near Side Megabasin, making it look somewhat like a pie with a large slice cut out (see Figure 13.3). This complex pattern of the two interacting basins probably contributed to the difficulty in perceiving it earlier. The ejecta pattern of the Near Side Megabasin is unlike that of any other basin, not because the dynamics of its transient crater and internal basin are any different, but simply because it is so large in relation to its spherical target. The manner of the deposit is shown in Figure 13.6.

Ejecta from Antipode to Escape Scarp

The impact that caused the Near Side Megabasin was so violent that much of the material ejected from the basin had escape velocity, and was lost to the Moon. When the ejection velocity fell below the escape velocity, it was still sufficient to throw the ejecta beyond the antipode. Then, as the explosion expanded and the velocity fell further, the material was deposited progressively back toward the antipode. A scarp (cliff or region of high slope) was left at the radius where the ejecta whose velocity was just short of the escape velocity fell.

In total, two layers of ejecta were deposited; one layer between the edge of the basin and the antipode and the other layer between the antipode back toward the edge. Although the ejecta forming the layer beyond the antipode left the basin first, it forms the upper layer, because, having a high velocity, it followed an orbit further out into space and landed after the layer between the rim and the antipode.

Antipode Bulge

The bulge on the far side of the Moon, shown clearly in the elevation map of Figures 13.2 and 13.3, was caused by the heavy double layer of ejecta that fell into the limited region near the antipode point. It must have formed an enormous mountain at the antipode. Alas, we cannot see that mountain now, because a subsequent impact formed the Korolev Basin (see Chapter 7). The antipode is well within that basin.

Model Compared to Measurement

The "proof of the pudding" of the pattern of ejecta caused by the Near Side Megabasin is a comparison of the model's estimate with the measured radial profile, centered at the impact point (see Figure 13.7). The vertical scale of the model in

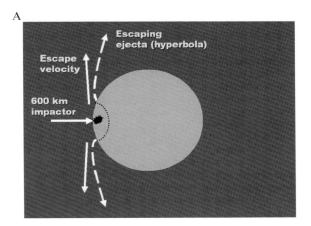

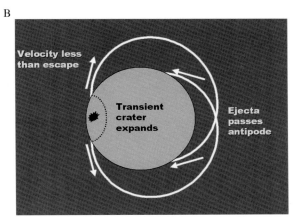

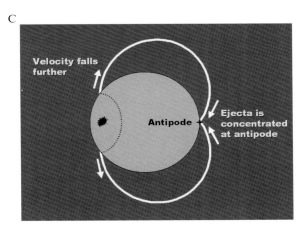

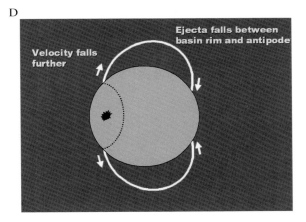

Figure 13.6. This series of sketches shows the evolution of the Near Side Megabasin as its transient crater expands. (**A**) The impactor of the Near Side Megabasin may have been 600 km in diameter, assuming that it was about one-tenth the basin diameter. The initial ejection velocity, appropriately scaled, would have been greater than the lunar escape velocity. (**B**) As the transient crater (the region of vaporized, molten, and fractured rock) expands, the velocity falls below escape velocity. Ejecta is deposited inside of a specific boundary that is nearly circular. The orbits are not to scale; their apolunes are much larger than shown. (**C**) As the transient crater continues to expand, the ejection velocity falls further and (at a specific time after impact) the ejecta all fall together at the antipode, producing the center of the far side bulge.(**D**) As the transient crater expands, the ejecta lands on the region between the antipode and the expanding transient crater. Ultimately, the two meet as the velocity falls to zero, at the rim of the Near Side Megabasin.

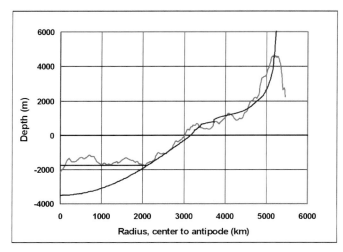

Figure 13.7. The radial profile of the Near Side Megabasin. The smoother curve is derived from the model of impact basins. The rougher curve is an average depth, as a function of radius, from the impact point at 8.5 north latitude and 22.0 east longitude. The measured profile excludes the southeast quadrant, where the South Pole-Aitken Basin lies.

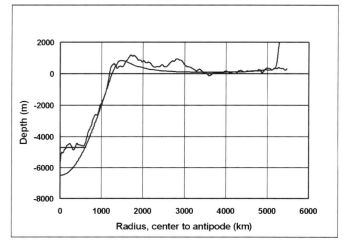

Figure 13.8. The radial profile of the South Pole-Aitken Basin, to the same scale as Figure 13.7. The measured profile excludes the southeast quadrant, where the Near Side Megabasin lies. The model of the South Pole-Aitken Basin indicates that the ejecta that had just less than escape velocity does not quite reach the antipode, although there is a large mound around the antipode. This mound cannot be directly observed today because it is beneath Mare Frigoris. The fill in the floor of this basin consists partly of mare and partly of ejecta from nearby basins.

Figure 13.7 is affected by isotopic adjustment in the same way as the elevation.

Isostasy has reduced all elevation variations produced in the crust during the early period when the crust and mantle were soft by a factor of 6 (see Section 13.5). At the time of its formation (before isostatic compensation), the Near Side Megabasin must have been 21 km deep at its center in order to produce enough ejecta, before it subsided due to isotropic adjustment.

Figure 13.7 shows a depression at the antipode (near the right side of the graph) that is due to the Korolev Basin. The level floor of the Near Side Megabasin reflects refilling of the basin from below. Ejecta from later basins lies on top of the refilled crust.

So when we look at the far side, nearly all that we see has been deposited there from the near side or is the effect of the South Pole-Aitken Basin or subsequent impacts. Only a few outbreaks of mare are truly native to the far side, coming from its lower crust or upper mantle. The geometry of the far side can be understood as evolving from the Near Side Megabasin; its rim, its ejecta blanket, the antipode mound, and the scarp resulting from the transition to ejecta returned to the Moon from ejecta sent into space.

13.4. The South Pole-Aitken Basin and its Effects on the Near Side

The South Pole-Aitken Basin was first identified from Lunar Orbiter photographs shot at an oblique angle. Later, the Clementine mission provided much more detail, revealing this basin to be the largest known up to that time, not only on the Moon but even in the entire Solar System. It is about 2400 km in diameter.

Model Compared to Measurement
The model and measured radial profile of the South Pole-Aitken Basin are shown in Figure 13.8. The model shown is

for a circular basin, although we know that the South Pole-Aitken basin is actually elliptical, with an eccentricity of about 0.7. Part of the discrepancy between the model and the measured radial profile is due to this elliptical shape.

Since the South Pole-Aitken Basin took a large bite out of the ejecta of the Near Side Megabasin, it must have come somewhat later. Like the earlier basin, it left one layer of ejecta over the whole Moon outside of its internal basin and a second layer near the antipode.

The model indicates that ejecta of the South Pole-Aitken Basin is currently at least 175 m in topography, 1050 m in crustal thickness, everywhere outside of its internal basin, including on top of the ejecta from the Near Side Megabasin. However, there appears to have been a refilling of the Near Side Megabasin (up to about 1700 m below the original surface). The mound at or near the antipode of the South Pole-Aitken Basin has been covered by a lava flow in the area called Mare Frigoris.

In a recent paper (Schultz, 2007), Peter Schultz has suggested that when the transient crater of the South Pole-Aitken Basin terminated, it launched a massive shock wave through the Moon. Because of the oblique velocity of the impactor, this shock wave would be offset from the antipode. It may have struck the northwestern near side, causing a fracture pattern that later resulted in ridges and faults in Oceanus Procellarum, Mare Imbrium, Mare Humorum, and other features, a pattern that has previously been reported by Schultz and others (Wilhelms, 1987).

13.5. Thickness of the Crust

The Moon, like Earth, has a crust formed of lighter minerals (anorthosites, mostly silicates) that have separated from the body of magma that resulted from the Moon's formation. The heavier minerals that were left are in the mantle, the material beneath the crust. Studies of the Moon's fine-scale gravitational field (Zuber, 1994; Lemoine, 1997), derived from tracking orbital spacecraft, together with topographic data, allow estimation of variations in thickness of the layer of crust (Hikida, 2007).

The gravity field over a topographic high (a mound) in the crust would increase due to the excess mass. When the gravity is smooth there, as has been observed over the highlands (Lemoine 1997), it is inferred that there is a compensating mass of crust depressed into the underlying mantle to support, through buoyancy, the extra weight of the mound.

The relative density of the crust and mantle material determines the ratio of the variation in total thickness of crust to the variation in surface elevation. For the values used in Hikida, 2007 (2.8 g cm^{-3} for crust and 3.36 g cm^{-3} for mantle), this ratio is 6.0. It follows that attributing the current topography of the far side to the variation in the deposit of ejecta from the Near Side Megabasin implies that the deposited depth is six times the current variation in elevation. Dividing the crustal thickness data provided by Hikida and Wieszorek by this value and setting the thickness of the pristine crust to 47.5 km provides an excellent match to the far side highlands, including the bulge.

To generate the ejecta needed to provide the amount that was added to the far side, the depth parameter of the Near Side Megabasin must be 21 km. Then, with the other parameters, and the model of the South Pole-Aitken Basin, there is a match between the ejecta and both topography and crustal thickness on the entire far side except within the South Pole-Aitken Basin.

On the near side, much of the crust that was originally there would be removed by the Near Side Megabasin impact. In addition to the transfer of the ejecta to the far side, a similar mass of crust would be driven off of the Moon. In addition, the crust under the basin may have been additionally thinned by some of it being driven deep into the mantle by the turbulence of the transient crater. Some crust probably was regenerated and rose into the cavity to form the level floor shown in Figure 13.7.

A large effect on the crust would be expected from the action of the transient craters below the Near Side Megabasin and the South Pole-Aitken Basin. They would extend downward for distances of the order of the radius of the basins, hundreds of kilometers. The turbulence within those transient craters may have driven crust down into the mantle, exchanging it for heavy mantle materials. The transient craters of subsequent large basins would reinforce this action. This could be the explanation of the measured thinness of the crust below the Near Side Megabasin and the South Pole-Aitken Basin.

Figure 13.9 shows how the Near Side Megabasin modified the thickness of the crust. The same effect would operate on the South Pole-Aitken Basin.

13.6. Parameters of the Near Side Megabasin

Table 13.1 contains the proposed parameters of the Near Side Megabasin. These parameters were adjusted to best fit both the topographic and crustal thickness data. The depth is that

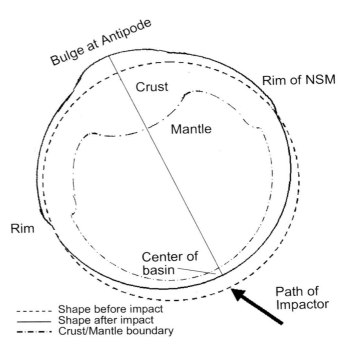

Figure 13.9. This cartoon shows how isostatic compensation affects the additional crust that is deposited by the Near Side Megabasin. It sinks down into the mantle, driving the primitive crust before it. The Near Side Megabasin has thinned the crust beneath its cavity by ejecting some of it and driving some of it deep into the mantle. Elevations and depths are exaggerated, relative to the radius of the Moon.

Latitude	8.5° north	Longitude	22° east
Major axis	3320 km	Minor axis	3013 km
Depth	21 km	Launch angle	50° from flat
Eccentricity	0.42	Orientation	53° west of north

Table 13.1. The Near Side Megasbasin

of the initial basin, before isostatic compensation and before regenerated crust filled the level floor. The current depth, including the regenerated crust and mare fill is about 1800 m from floor to the estimated original target surface.

13.7. Asymmetry of the Center of Mass

The transfer of crust from the near side to the far side would have moved the center of mass backward relative to its initial position. But since the crust is lighter than the mantle, the center of figure would be shifted back even more, so that the center of mass would have a net motion forward, as is observed.

13.8. Heavy Element Anomalies

The high concentrations of heavy elements (iron, magnesium, and thorium) that are observed within the South Pole-Aitken Basin and the proposed Near Side Megabasin may have been stirred up by the turbulence of the fractured, molten, and vaporized material in the transient crater (Ivanov, 2007). As it drove crust down, it would have brought up mantle material.

13.9. Summary

The following hypothetical scenario accounts for both the topography and the thickness of the crust, in shape and in magnitude.

1. The primitive crust was 47.5 km deep and uniform.
2. The Near Side Megabasin formed a cavity (21 km deep in the center) over the near side of the Moon, extending beyond the limbs. Below the cavity, crust was driven into the mantle and mantle material raised. Some ejecta escaped from the Moon and the rest covered the far side, with a depth at the bulge of tens of kilometers.
3. The ejected crust subsided, carrying the primitive crust down into the mantle. The ejecta left above the mean surface (about 1/6 of the original depth) established the current topography of the far side. The cavity of the Near Side Megabasin had a similar compression.
4. The South Pole-Aitken Basin was formed after the Near Side Megabasin, with a vertical scale 6 times its current scale. It too experienced isostatic compensation.
5. The crust beneath the two megabasins partly returned from the transient craters, leaving the level floors that are found today.

An important question is "How well does our model compare with the measured topography of the Moon?" To answer this question, parameters of over 50 large basins as they are today, including the two giant basins following isostatic compensation, were estimated. Their internal cavity, internal rings, ejecta blanket, and mare or crustal fills were simulated. The resulting comprehensive model is shown in Figure 13.10, along with the current elevation map of Figure 13.2.

As is readily seen, the model and the elevation map are in close qualitative and quantitative agreement, much as is the radial profile of Figure 13.7.

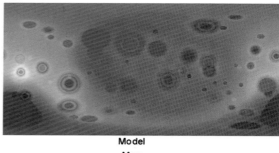

Model
Map

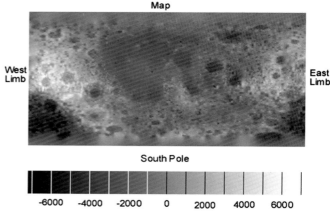

West Limb

East Limb

South Pole

-6000 -4000 -2000 0 2000 4000 6000

Elevation (m)

Figure 13.10. The bottom figure is the map of the current elevations of the Moon, as measured by the LIDAR instrument of Clementine. The upper figure is a simulation of the Near Side Megabasin, the South Pole-Aitken Megabasin, and about 50 smaller basins and large craters. Note the strong qualitative and quantitative agreement, especially for the larger features.

Glossary

Albedo: The inherent brightness of a surface. The Moon is much brighter when the sun is directly behind the observer, which is why the full Moon is so bright.

Anorthosite: A type of rock composed of relatively light-colored, low density silicate minerals.

Antipode: The point directly on the other side of the Moon from the center of an impact crater.

Apparent crater: The present surface cavity of an impact feature.

Basalt: Rock containing a collection of minerals relatively low in silica, containing iron and magnesium, similar to the floors of Earth's oceans.

Basin: Large craters have distinctive features, internal and external rings in particular, which smaller craters lack. By convention, impact features with rim diameters of more than 300 km are called basins.

Catena: A linear feature composed of a chain of craters. These are usually secondary craters, forming a chain radial to their parent crater or basin.

Copernican Period: The period from the end of the Eratosthenian Period to the present. The crater Copernicus is the archetype feature of this period.

Crater: A compact depression. Nearly all craters on the Moon are caused by impact from asteroids or comets. Craters have raised rims and are surrounded by ejecta blankets.

Crust: The upper layer of rock composed of lighter minerals such as silicates that have separated from a melt and risen above the mantle because of their low density. The surface of the Moon is crust, except where it is covered by maria. Since the maria pool in the cavities of impact features, the areas of exposed crust tend to be higher and are often called highlands.

Cryptomare: It is sometimes possible to infer the existence of the dark flat surface of basaltic mare even though it is hidden by bright debris from surrounding impact features.

Early Imbrian Period: The age range from the Imbrium Basin event to the Orientale Basin Event. This period is often called an epoch.

Ejecta: The material thrown out from the cavity of an impact crater. Some ejecta from lunar impact features reached lunar escape velocity and was lost to the Moon. Meteoroids that have struck Earth and been recovered include some ejected from craters of the Moon and of Mars.

Ejecta blanket: The continuous layer of ejecta extending for approximately one crater (or basin) radius outside of the rim of an impact crater or basin.

Eratosthenian Period: The age range between the end of the Late Imbrian Period and the start of the Copernican Period. The crater Eratosthenes is the archetype of this period.

Geoid: A surface of equal gravitational potential around a planet or moon, approximately an oblate spheroid, due to rotation of the body. The reference geoid of the Moon is at a mean radius of 1738 km.

Impactor: A body that strikes a target surface and forms an impact crater there. Impactors usually come from the asteroid belt, but they may be comets from the Kuiper belt.

Isostasy, isostatic compensation: A condition of balance between the weight of a body and the buoyant force of a supporting medium that lacks shear strength. When the mantle was still liquid or plastic at the time of the Magma Ocean, a block of crust of a given thickness would settle into the mantle until only a sixth of its depth was above the mean surface.

KREEP: An acronym for a mineral consisting of potassium (Chemical symbol K), phosphorus (P), and the collection of elements called rare earth elements (REE). These REE elements are sometimes called "incompatible" because they are the last to solidify from a melt as it cool and the first to melt as a solid heats. Since these elements melt early, a mix of them tends to accompany rising lava.

Hypervelocity Impact: An event caused by an impactor traveling at a speed relative to the target surface that exceeds the speed of sound in that surface.

Lacus: A small depression filled with lava flows.

Late Imbrian Period: The age range between the Orientale Basin event and the start of the Eratosthenian Period. Most maria formed in this period. This period is often called an epoch.

LIDAR: Laser Image Detection and Ranging instrument carried on the Clementine spacecraft. A laser altimeter whose measurements were combined with spacecraft orbital data to provide an elevation map of the Moon.

Limb: An edge of the visible disk of a body such as the Moon, especially a semi-circle from pole to pole. The East Lmb of the Moon as seen from Earth is at 90° east longitude and the West Limb is at 90° west longitude. If the Moon is viewed from the far side, the directions are reversed; the limb at 90° east is toward the west, but it is still called the East limb so that we can say "Mare Smythii is at the East Limb" no matter what our viewpoint.

Magma Ocean: The analysis of lunar samples has led to a consensus that the minerals of the crust (highlands) and maria separated from molten rock (magma). An ocean of magma formed when the Moon coalesced from the vapor, melt, and debris left over after the collision with Earth.

Main ring: The highest circular ridge that surrounds the cavity of a basin (also called the rim).

Mantle: The layer of rock below the crust and above the core. It is composed of minerals that have separated from a melt and fallen because of their high density.

Mare, maria: Many depressions, especially on the near side of the Moon, are filled with basalt that has erupted from below. This dark material was deposited at a low viscosity because of its low volitile content. The maria form the patterns we see on the near side of the Moon.

Mascon: A concentrated mass that affects the gravity field of the Moon. Mascons are associated with pipes of high-density mantle material that flooded maria.

Mass-wasting: The steady rain of small meteoroids onto the surface of the Moon erodes boulders and sharp features of the topography. The degree of mass-wasting by this process is an indication of age of a feature.

Megabasin: A basin that contains smaller basins within it.

NSM: Near Side Megabasin.

Near Side Megabasin: This is the giant impact basin on the near side that I propose threw its ejecta on the far side very early in the history of the Moon. See Chapter 13.

Nectarian Period: The age range from the Nectarian Basin event to the Imbrium Basin event.

Pre-Nectarian Period: The age range prior to the formation of the Nectarian Basin.

Plains: Smooth areas on the lunar surface, perhaps caused by the deposit of finely divided ejecta from an impact feature beyond its ejecta blanket. Dark plains are probably caused by fire-fountains, volcanic features that emit beads of glass propelled by gas emissions.

Rays: Young impact features show bright star-like patterns called rays spreading radially for very long distances, far beyond ejecta blankets. Rays are characterized as optical or compositional. Compositional rays are composed of brighter minerals than the background strata. Optical rays have formed too recently to be darkened by exposure to the solar wind.

Rima: A long, narrow depression.

Rings: A circular ridge that is formed along with a basin. The rim is called the main ring, internal rings are inside the main ring, and external rings are outside the main ring.

Ringed Basins: See "Basin".

Secondary crater: A crater formed by the impact of material ejected from another primary crater or basin.

Theia: The small planet that impacted Earth to form the Moon. An alternate name is Orpheus.

SPA: South Pole-Aitken Basin.

Stratigraphy: The column of successive layers of material deposited by ever-younger features according to the principle of superposition.

Superposition: A younger feature usually lies above an older feature and adds its depth profile to the surface formed by the features below. To understand the shape of a surface feature, the shape of the substrate needs to be inferred. Sometimes the shape of an older feature can be seen by its perturbation of the younger feature above it, especially older craters under the ejecta blanket of younger craters or basins.

Target surface: The surface that was struck by an impactor to form an impact feature. The characteristics of the resulting feature depend on properties of the target surface.

Topography: The shape of a surface.

Transient crater: An impact crater produces a roughly hemispherical region below it of fractured, melted, and vaporized minerals. The agitated condition lasts for only a short time: that is the reason for the term. A transient crater has a much greater volume than the visible cavity on the surface.

Vallis: A wide long depression.

References

Bowker, 1971: D.E. Bowker and J.K. Hughes, *Lunar Orbiter Photographic Atlas of the Moon*, NASA SP 206.

Bussey, 2004: B. Bussey and P.D. Spudis, *The Clementine Atlas of the Moon*, Cambridge University Press, Cambridge.

Byrne, 2002: C.J. Byrne, *Automated cosmetic improvements of mosaics from Lunar Orbiter*, LPSC 2002, Lunar and Planetary Institute, Houston (CDROM).

Byrne, 2005: C.J. Byrne, *Lunar Orbiter Photographic Atlas of the Near Side of the Moon*, Springer-Verlag, London.

Byrne, 2006: C.J. Byrne, *The Near Side Megabasin of the Moon*, Abstract # 1930, LPSC 2006, Lunar and Planetary Institute, Houston (CDROM).

Byrne, 2007: C.J. Byrne, *Interior of the Near Side Megabasin of the Moon*, Abstract # 1248, LPSC 2006, Lunar and Planetary Institute, Houston (CDROM).

Cadogan, 1974: P.H. Cadogan, *Oldest and largest lunar basin?* Nature 250:315–316.

Feldman, 2002: W.C. Feldman, et al., *Global distribution of lunar composition: New results from Lunar Prospector*, Journal of Geophysica Research 107(E3).

Gaddis, 2001: L. Gaddis, et al., *Cartographic processing of digital zlunar Orbiter data*, LPSC 2001, Abstract #1892, Lunar and Planetary Institute, Houston.

Garrick-Bethell, 2004: I. Garrick-Bethell, *Ellipses of the South Pole-Aitken Basin: Implications for basin formation*, LPSC 2004, Abstract 1515.

Gillis, 2002: J. Gillis, Editor, *Digital Lunar Orbiter photographic atlas of the Moon*, www.lpi.usra.edu/research/lunarorbiter, Lunar and Planetary Institute.

Hawke, 2007: B.R. Hawke et al., *Remote sensing studies of Copernican Rays: Implications for the Copernican - Eratosthenian boundary*, LPSC 2007, Abstract #1133, Lunar and Planetary Institute, Houston (CDROM).

Hikida, 2007: H. Hikida and M.A. Wieczorek, *Crustal thickness of the Moon: New constraints from gravity inversions using polyhedral shape models*, LPSC 2007, Abstract #1547, Lunar and Planetary Institute, Houston (CDROM).

Ivanov, 2007: B.A. Ivanov, *Lunar impact basins – Numerical modeling*, LPSC 2007, Abstract #2003, Lunar and Planetary Institute, Houston (CDROM).

Lemoine, 1997: F.G.R. Lemoine et al., *A 70th degree lunar gravity model (GLGM-2) from Clementine and other tracking data*, Journal of Geophysica Research 102(E7).

Prettyman, 2007: T.H. Prettyman, et al., *Analysis of low-altitude Lunar Prospector gamma ray spectra*, LPSC 2007, Abstract #2214.

Schultz, 2007: P.H. Schultz, *A Possible Link Between Procellarum and the South Pole-Aitken Basin*, LPSC 2007, Abstract #1839, Lunar and Planetary Institute, Houston (CDROM).

Spudis, 1993: P.D. Spudis, *The Geology of Multi-Ring Impact Basins: The Moon and Other Planets*, Cambridge University Press, Cambridge.

Stevens, 1999: R. Stevens, *Visual Basic Graphics Programming, Second Edition*, Wiley, New York.

Whitaker, 1981: E.A. Whitaker, *The lunar Procellarum Basin*, Multi-ring basins, LPSCP 12, Part A.

Whitaker, 1999: E.A. Whitaker, *Mapping and Naming the Moon*, Cambridge University Press, Cambridge.

Wilhelms, 1987: D.E. Wilhelms, et. al., *The Geologic History of the Moon*, USGS Professional Paper 1348, US Government Printing Office, Washington.

Wood, 1973: J.A. Wood, *Bombardment as a Cause of the Lunar Asymmetry*, The Moon 8, 1973.

Zuber, 1994: M.T. Zuber, D.E. Smith, and G.A. Neumann, *The shape and internal structure of the Moon from the Clementine Mission*, Science, 266:1839–1843.

Zuber, 2004: M.T. Zuber, D.E. Smith, and G.A. Neumann, *topogrd2*, web site of the Univeristy of Washington at St. Louis.

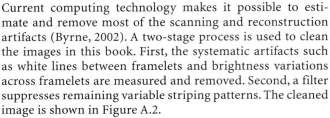

Appendix A
Cleaning the Scanning Artifacts from Lunar Orbiter Photos

A.1. Nature of the scanning artifacts

An example of a typical Lunar Obiter subframe, as downloaded from the web page of the Lunar and Planetary Institute, is shown in Figure A.1.

The scanning artifacts ("venetian blind effect") make it difficult to visualize the lunar surface.

A.2. A program to clean the artifacts

Current computing technology makes it possible to estimate and remove most of the scanning and reconstruction artifacts (Byrne, 2002). A two-stage process is used to clean the images in this book. First, the systematic artifacts such as white lines between framelets and brightness variations across framelets are measured and removed. Second, a filter suppresses remaining variable striping patterns. The cleaned image is shown in Figure A.2.

The software programs for each of the stages were written in Visual Basic® and run in a Windows 98 environment. They use utility routines from "Visual Basic Graphics® Programming, a book by Rod Stevens (Stevens, 1999).

Steps of Stage 1

The steps performed by the program that measures and removes systematic artifacts (Stage 1) are the following:

1. Search from the top of the mosaic image for the first identifiable bright line between framelets (a framelet edge)
2. Jump half a typical 40-pixel framelet width and search for the next identifiable framelet edge
3. Repeat the jump-and-search process until the bottom of the image is reached
4. Calculate the precise framelet width of the image (which varies between 38 and 43 pixels on the currently sampled images) by averaging the framelet width between successive identified framelet edges. Successive edges are those separated by about one typical framelet width

5. Extrapolate or interpolate those framelet edges that were not identified by search processes
6. Scan each framelet edge and, for each horizontal pixel coordinate, estimate the excess brightness by comparison with the immediately adjacent pixels. Then remove the excess brightness from the pixels of the affected scan lines
7. Determine the average linearized brightness ratio of the pixels, as a function of their relative position between framelet edges, for the entire mosaic image. A model of the nonlinear contrast function is used in the linearization process
8. Correct the brightness of each pixel for the average normalized brightness ratio corresponding to its position between framelet edges. The contrast model is used in the correction.

Specific Techniques of Stage 1

The program addresses several problems that are specific to the mosaics.

Finding the edges of framelets. Usually, framelet edges are represented by bright lines caused by light shining between the framelet filmstrips as the mosaics were laid up. Although they are narrower than the distance between scans of the digital images, the lines appear in either two or three adjacent lines of pixels because the scan spot used in creating the digital images was larger than the distance between scans. The scan lines are usually tilted with respect to the framelet edges by two or more pixels across a frame and have curvature of the order of one pixel across a frame. In many frames, the white lines are obscured by saturation or overcome by image signal in large parts of the frame. Development artifacts and some valid signals can produce false segments of bright lines that are not at framelet edges.

The approach taken to find these lines is to search in a band of scan lines 17 pixels high, looking in 17×21 blocks of pixels for a bright white line segment of either two or three pixels in width. A quadratic best-fit line is calculated using the centers of the line segments. Successive fits are calculated, eliminating those centers that are most off the line until all remaining centers are within half a pixel of the line. If the remaining centers are at least 25% of the possible centers, the

Figure A.1. LO4-187H2, Mare Orientale and Montes Rook.

Figure A.2. LO4-187H2 after removal of systematic artifacts and filtering.

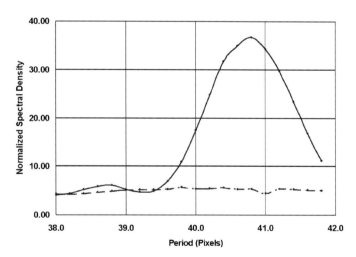

Figure A.3. Spectral density of LO4-140H3 for periods near the framelet width (about 40 pixels). The *solid line* shows the spectrum of the raw photo and the *dashed line* shows the spectrum after removal of the systematic artifacts.

line is accepted as a framelet edge. If not, the band is moved down the picture to find an edge.

Removing each framelet edge line. For each horizontal pixel index, the pixels above and below the calculated vertical value of the best-fit line are examined to determine the probable value of the excess brightness of the line and a weighted value is subtracted from the brightness of the two or three pixels near the line. This process preserves more detail in the image than simply averaging the brightness around the line.

Contrast model. The streaking artifact was applied in the spacecraft on a low contrast image, essentially linear in its relation between photographed brightness and density.

However, the atlas images were printed at high contrast and are therefore nonlinear. Thus, in order to determine and compensate for streaking, it is necessary to linearize the brightness by reversing the contrast function. An empirical study was conducted, varying the assumed contrast function and minimizing the residual fundamental and second and third harmonics of the artifacts at the framelet frequency. The resulting contrast function has a linear contrast gain of 3 between output brightness values of 0.1–0.9 of the full range and exponential curves at each end of the brightness range.

Special handling. A few images require special techniques. For example, a feature has been added to compensate for an alternating pattern of darker and lighter framelets in some of the images.

Quantitative Results of Stage 1

A spectral analysis program has been written to compare the spectra of the input and output images to provide a quantitative measure of improvement. Figure A.1 shows the fine-scale normalized vertical spectral density of LO4-140H3 before and after processing.

The removal of systematic artifacts greatly improved the appearance of the pictures. The venetian blind effect disappears, providing a subjective effect of more direct visual contact with the lunar surface. However, traces of the framelet artifacts remain in parts of some of the images.

Steps of Stage 2

Stage 2 is a two-dimensional filter process that performs the following steps:

1. Generate the two-dimensional Fourier transform of the image

2. Multiply the transform by a two-dimensional filter that suppresses streaks, that is, patterns that extend horizontally more than vertically. This filter is formed from the combination of a low-pass filter and a high-pass filter, as suggested by Lisa Gaddis of USGS (Gaddis, 2001).
3. Return to the spatial domain by an inverse two-dimensional Fourier transform.

The remaining artifacts are substantially removed by application of the two dimensional filter (see Figure A.2). Careful comparison of filtered images with the originals verifies that shadows, topographic variations, and albedo variations such as associated with rays are preserved. This is true even when the digital images are examined under magnification that resolves the pixels.

Specific Techniques of Stage 2

Fourier Transform and inverse transform: A fast-Fourier-transform algorithm runs in the vertical direction and stores arrays of the real and imaginary parts of the transform. The same algorithm runs in the horizontal direction on these arrays to create the real and imaginary two-dimensional arrays that are now in the frequency domain. The same process is used for the inverse transform. The fast-Fourier-transform algorithm used is limited to image arrays that have dimensions M and N that must be integer powers of two; therefore, the dimensions of the actual images are built up to the next power of two by adding virtual blank brightness data.

Filter characteristic: The filter has a low-pass characteristic in the vertical (north–south) direction to suppress the framelet frequency and its harmonics. In the horizontal direction (east–west), it has a high-pass characteristic to reject patches of horizontal stripes while permitting topographic detail to remain.

Both filters are of the second-order Darlington type.

A second-order low-pass Darlington filter has the form

$$G(f) = \frac{1}{1 + (fT)^2},$$

where f is the frequency and T is the period of the frequency that reduces the filter function $G(f)$ to 0.5.

A high-pass Darlington filter has the form

$$G(f) = 1 - \frac{1}{1 + (fT)^2}.$$

Examination of the two-dimensional Fourier transform of an input image (whose systematic artifacts have been removed) shows strong noise at horizontal frequencies whose period is less than 150 pixels and vertical frequencies whose period is more than 25 pixels.

A compound (two-dimensional) "Normal" filter with the following form is used to suppress the noise:

$$G(fx, fy) = 1 - \left\{ \frac{1}{1 + (fx \times 150)^2} \left(1 - \frac{1}{1 + (fy \times 25)^2} \right) \right\},$$

where fx is the frequency in the x direction (east–west) and fy is the frequency in the y direction (north–south).

The "Normal" filter is used as the default, and is effective for most of the pictures. If streaks remain after application of the "Normal" filter, a "Strong" filter is used instead. This happens most often when large areas of dark sky or shadow beyond the terminator appear in an image. The form of the "Strong" filter is

$$G(fx, fy) = 1 - \left\{ \frac{1}{1 + (fx \times 50)^2} \left(1 - \frac{1}{1 + (fy \times 50)^2} \right) \right\}.$$

A.3. Summary

The Lunar Orbiter photographs have been used as the major reference for topographic information for the last 40 years. They commonly appear in papers about the Moon even today in their original state, artifacts included.

Throughout this time, researchers have learned to distinguish the artifacts from the topographic information. It is my hope that these cleaned photographs will make the nature of the Moon more available to both nonprofessionals and to new students of the Moon.

General Index

This **General Index** covers Chapters 1 through 5, Chapter 13, the Appendix, and the introductory pages of each chapter of regional coverage. The features in the comprehensive photographs covering each region are listed in the **Index of Annotated Features**. Some small features that do not appear explicitly in the photographs are listed in the **Index of Far Side Features that are not Annotated**. Apollo, Clementine, and Lunar Orbiter photos are listed in the **Index of Photographs**.

Index of Annotated Features

This **Feature Index** includes all annotated features in the regional chapters (5 through 12). Additional entries for some features can be found in the **General Index**.

Index of Far Side Features that are not Annotated

The following small features, each with a diameter less than 50 km, are not annotated on the photos in this book. They are listed here for completeness and to assist in locating them if so desired. The page given is the best photo containing the feature. A nearby feature that is annotated is identified for each of these small features and a page showing that feature is listed. The names in **bold** print are explicitly labeled on the Lunar Aeronautical Charts (LAC). The column marked "LAC" is the page number in "The Clementine Atlas of the Moon" (Busey, 2004). Some of these features were formerly designated by letter codes and are shown as such on the LAC charts.

Name	Diam. (km)	Latitude	Longitude	Page	Formerly	Nearby Feature	LAC
Andersson	13	48.7 S	95.3 W	178		De Roy	135
Balandin	12	18.9 S	152.6 E	80		Grave	103
Bawa	1	25.3 S	102.6 E	23		Bowditch	100
Bergstrand	43	18.8 S	176.3 E	80		Aitken	104
Bondarenko	30	17.8 S	136.3 E	25	Patsaev G	Patsaev	102
Chadwick	30	52.7 S	101.3 W	178	De Roy X	De Roy	135
Chalonge	30	21.2 S	117.3 W	158		Lewis	107
Chang Heng	43	19.0 N	112.2 E	97		Olcott	47
Coblentz	33	37.9 S	126.1 E	107		Tesla	30
Ctesibius	36	0.8 N	118.7 E	25		Hero	65
Debus	20	10.5 S	99.6 E	23	Ganskiy H	Hansky	82
Edith	8	25.8 S	102.3 E	23		Bowditch	100
Ewen	3	7.7 N	121.4 E	25		Ibn Firnas	65
Fairouz	3	26.1 S	102.9 E	23		Bowditch	100
Feoktistov	23	30.1 N	140.7 E	98		Tereshkova	48
Fesenkov	35	23.2 S	135.1 E	26		Stark	102
Ginzel	35	14.3 N	97.4 E	22		Mobius	64
Hayford	27	12.7 N	176.4 W	29		Virtanen	68
Heyrovsky	16	39.6 S	95.3 W	178	Drude S	Drude	123
Hohmann	16	17.9 S	94.1 W	176		Mare Orientale	108
Hume	23	4.7 S	90.4 E	23		Hirayama	82
Il'in	13	17.8 S	97.5 W	176		Mare Orientale	108
Ingalls	37	26.4 N	153.1 W	130		Joule	51
Karima	3	25.9 S	103.0 E	23		Bowditch	100
Kimura	28	57.1 S	118.4 E	68		Fechner	130
Kira	3	17.6 S	132.8 E	20		Patsaev	102
Konoplev	25	28.5 S	125.5 W	152	Ellerman Q	Von der Pahlen	107
Kosberg	15	20.2 S	149.6 E	28		Gagarin	102
Kramerov	20	2.3 S	98.8 W	150	Lenz K	Grachev	90
Krylov	49	35.6 N	165.8 W	102		Schneller	33
Leonov	33	19.0 N	148.2 E	98		Belyaev	48
Marci	25	22.6 N	167.0 W	102		Jackson	51
McAdie	45	2.1 N	92.1 E	23		Babcock	64
Melissa	18	8.1 N	121.8 E	25		Ibn Firnas	65
Mons Andre	10	5.2 N	120.6 E	25		King	65
Mons Dieter	20	5.0 N	120.2 E	25		King	65
Mons Dilip	2	5.6 N	120.8 E	25		King	65
Mons Ganau	14	4.8 N	120.6 E	25		King	65
Murakami	45	24.3 S	140.5 W	152	Mariotte Y	Das	106
Parsons	40	37.3 N	171.2 W	102		Schneller	33
Pikel'ner	47	47.9 S	123.3 E	26		Kozyrev	117
Ravi	2.5	22.5 S	151.8 E	28		Raspletin	103
Romeo	8	7.5 N	122.6 E	25		Ibn Firnas	65
Safarik	27	10.6 N	176.9 E	48		Sharanov	68
Shahinaz	15	7.5 N	122.4 E	25		Ibn Firnas	65
Soddy	42	0.4 N	121.8 E	25		Hero	65
Swann	42	52.0 N	112.7 E	96		Compton	16
Tsu Chung-Chi	28	17.3 N	145.1 E	98		Belyaev	48
Van Gent	43	15.4 N	160.4 E	29		Spencer Jones	67
Van Rhijn	46	52.6 N	146.4 E	166		Stormer	17
Wiechert	41	84.5 S	165.0 E	76		Shoemaker	144
Winkler	22	42.2 N	179.0 W	101		Duner	33
Zasyadko	11	3.9 N	94.2 E	23		Babcock	64

Index of Photographs

These are the photos and images that are printed in this book. All of the Lunar Orbiter photos of the far side that are in the Bowker and Hughes book (Bowker, 1971) are in the enclosed compact disk (CD) and are listed there, but a few of these do not appear on the printed pages and are not listed here.

Printed in China